FIRE
STRIKE
7/9

Dorset Libraries
Withdrawn Stock

For

Corporal Paul 'Sandy' Sandford and Guardsman Daryl Hickey,
gone but not forgotten

And for all the British and Allied soldiers who have lost their lives
in the ongoing conflict in Afghanistan

Honour the fallen

FIRE STRIKE 7/9

SERGEANT PAUL 'BOMMER' GRAHAME

AND DAMIEN LEWIS

EBURY
PRESS

1 3 5 7 9 10 8 6 4 2

First published in 2010 by Ebury Press, an imprint of Ebury Publishing
A Random House Group company
This edition published 2011

Copyright © Damien Lewis and Paul Grahame 2010

Damien Lewis and Paul Grahame have asserted their right to be identified as the
author of this work in accordance with the Copyright, Designs and Patents Act 1988

All photos © Paul Grahame unless otherwise stated

The Random House Group Limited Reg. No. 954009

Addresses for companies within the Random House Group can be found at
www.randomhouse.co.uk

A CIP catalogue record for this book is available from the British Library

The Random House Group Limited supports The Forest Stewardship Council (FSC),
the leading international forest certification organisation. All our titles that are
printed on Greenpeace approved FSC certified paper carry the FSC logo. Our paper
procurement policy can be found at www.randomhousebooks.co.uk/environment

Mixed Sources
Product group from well-managed
forests and other controlled sources
www.fsc.org Cert no. TT-COC-2139
© 1996 Forest Stewardship Council

Designed and set by seagulls.net

Printed in the UK by CPI Cox & Wyman, Reading, RG1 8EX

ISBN 9780091938086

To buy books by your favourite authors and register for offers visit
www.randomhousebooks.co.uk

CONTENTS

'*Merebimur* – We shall be worthy'
'*Viret in Aeternum* – It flourishes for ever'
Regimental mottos of The Light Dragoons

'Stand Firm and Strike Hard'
Regimental motto of 2 MERCIAN

ACKNOWLEDGEMENTS

Very special thanks to my then commanding officer at The Light Dragoons, Colonel Angus Watson MBE, for his invaluable support during the conception and writing of this book. Thanks to Captain James Kayll, also of The Light Dragoons, for his commitment and help in bringing this book to fruition, and to Major Antony Pearce, for all his friendship and support. My gratitude to all at my regiment, The Light Dragoons, for the comradeship and unstinting support over the years, and especially Sergeant Grant Cuthbertson, for getting me combat-ready in time for my Afghan deployment. All the staff at JFACTSU (Joint Forward Air Control Training and Standards Unit) deserve a very special mention, for their hard work in training me and others for the Afghan theatre. Very special thanks are due to Captain Chris Lane, Lance Bombardier Martin Hemmingfield, Lance Bombardier Ben Stickland and Bombardier Karl Jessop, my teammates in my Fire Support Team during our Helmand deployment. Without your help and enthusiastic support this book would not have been possible. Very special thanks also to Lieutenant Colonel Simon Butt, of 2 MERCIAN, the Officer Commanding (OC) of the unit in which I was embedded in Afghanistan. Your single-minded support and determination to help see this book through to fruition has been a battle winner in every sense. Very special thanks also to Major Stewart Hill, also of 2 MERCIAN, Commanding Officer during the last weeks of our Helmand tour, for your invaluable support and help. My gratitude also to Warrant Officer 2 Jason Peach of 2 MERCIAN, who was a tough and inspirational soldier to all, Sergeant Danny Fitzgerald of 2 MERCIAN, who did his utmost to make my job easier during the worst contacts imaginable and to all the soldiers of 2 MERCIAN whose commitment to win is exceptional. A note of thanks is also due to The Light Dragoons JTACs (Joint Terminal Attack Controller) presently in training and deployment, including Nick, Stu, Aldo and Wardy, for all your support. Further thanks are due to the JTACs who were deployed alongside me in Afghanistan, including Spunky, Dave, Reg, Chris, Si, Jamie, Stu, Bradders and Damo, who all had a hard and testing tour.

Special thanks also to our agent, Annabel Merullo, and to Tom Williams who works alongside her, for a guiding hand during the publishing process. Special thanks to our publisher, Andrew Goodfellow, and to all who worked on the lightning-fast production of this book – including Liz Marvin, Sarah Bennie, Caroline Craig and Alex Young. Thanks again to Alan Trafford, for his early reading of the manuscript. Sincere thanks also to David Beamont and Paula Edwards at the Ministry of Defence, and to Tim David, at HQ Land Forces, for your deft touch in helping to bring this book through to publication.

I would also like to thank my family, for all your support over the years – and for looking after me from the start (I know it wasn't always easy). Finally, my very special love and gratitude as always to Nicola, Harry and Ella for your support and patience during the process of making this book happen. I could not have done it without you.

AUTHOR'S NOTE

During my six-month deployment to Afghanistan I was part of a Fire Support Team that included four fellow soldiers. Our team call sign was *Opal Five Eight*. As the Joint Terminal Attack Controller (JTAC) within *Opal Five Eight*, my baby was the air power – calling in the ground attack aircraft, fast jets, helicopter gunships and unmanned aerial vehicles that perform such a vital role in modern warfare. My story as told in this book concerns my tour as I fought it. Doubtless, calling in and controlling the fire of the field guns – the sister role of *Opal Five Eight* – played an equally important part in the battles we fought, but this is the story of my war, the lightning-fast air-to-ground battle.

I could not have done what I did as a JTAC without the support of my fellow warriors in *Opal Five Eight*, and I hope that is clear from the pages that follow. I have done my very best to ensure that all the events portrayed herein are factually accurate, and to portray the battles we fought realistically and as they happened. No doubt my memory, and that of my fellow soldiers, is fallible, and I will be happy to correct any inadvertent mistakes in future editions.

The high level of ordnance dropped was specific to the battle environment and bound to the threat we encountered during my deployment. Since then battle conditions have altered and new directives on the use of air support, such as the COMISAF directive, have meant that such a high level of ordnance usage is no longer the norm.

In the latter weeks of my Afghan tour, the regiment that I was embedded with, 1 Worcester and Sherwood Foresters (1 WFR), was amalgamated into 2 MERCIAN Regiment. I have used the name

2 MERCIAN throughout this book, for simplicity's sake, and because that is the name by which the regiment is now known.

We have made every effort to identify the copyright holder of all the photographs used herein, and credit them accordingly. If there are any mistakes or errors in doing so, we will happily correct them in future editions.

The body count from my Afghan tour is an unofficial estimate based on my own JTAC log, the records of the pilots with whom I operated, and the B Company 2 MERCIAN combat diary. It has not been confirmed by the Ministry of Defence.

A donation will be made by the authors from the proceeds of this book to: Tickets for Troops, Combat Stress, Help for Heroes, The Light Dragoons Charitable Trust and 2 MERCIAN Regimental Benevolent Fund.

'We fought and died together in Afghanistan. The ethos was look after your mate and your mate will look after you, and we built it up from there. We dropped danger-close dozens of times. That's not how we trained for it, but that's how it had to be done on the ground, to win battles. And to say that we did those danger-close drops at night dozens of times and had no casualties – that says how effective the ground-to-air relationship was. At the end you've all trusted each other with your lives and formed an unbreakable bond – there's nothing ever that will come close to that.'

Major (now Lieutenant Colonel) Simon Butt, Officer Commanding, B Company 1 WFR (now 2 MERCIAN)

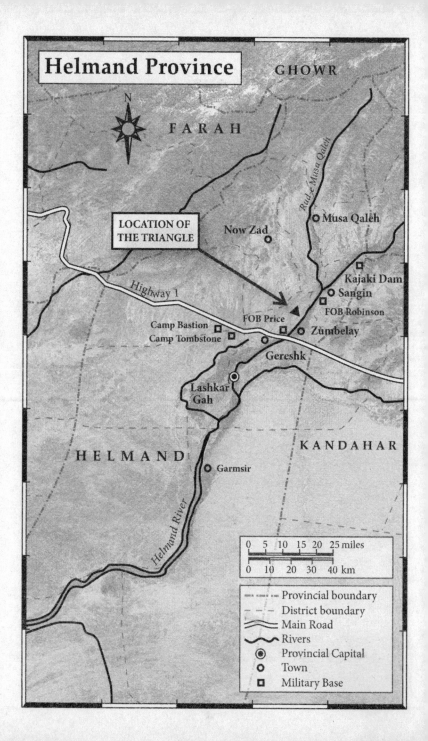

ONE
LIVE TO KILL

There it was again. Movement. A smudge of black amongst the shadows of the woodline. The shape of a male fighter? The flicker of a turban or a robe? Or just an animal?

Sticky nudged me in the ribs. 'You see it? Black flash between two trees. Wood strip, two hundred metres due east on the demarcation line.'

I nodded. 'Aye, I see it.'

I flicked my eyes across to the wagon. Chris was jammed in the Vector's armoured turret. His mop of blond hair was sticking out at all angles from under his helmet, his eyes searching everywhere for the enemy. The hulking great figure of Throp was hunched over the wheel, ready to gun the engine if we got targeted.

We sure as hell were going to, sticking out like dog's bollocks up here on the ridge. But where else were we to position ourselves? From here we had perfect visibility. The Green Zone rolled out before us, a lush carpet of vegetation bathed in the golden light of dawn. Here and there a line of trees traced a watercourse or a dirt road.

It looked deceptively peaceful. But this was bandit country, stuffed full of six hundred battle-hardened Chechen and Pakistani insurgents, or so the Intel boys had told us. It was 0630, and directly below us one hundred lads from B Company 2 MERCIAN were about to advance on foot into the jungle and the enemy guns. They were outnumbered six to one, and relying on us to even up the odds a little.

Stuck up here on this desert ridge we were the proverbial bullet magnet, devoid of any cover. But this was the place to get eyes on the battlefield, and call in air power to smash the enemy. We were pushing into their stronghold. Kicking the hornets' nest. The odds were stacked against us. Air power was about all we had to even things up a little.

I eyed that treeline again. 'Sticky, what d'you reckon ...'

I didn't get to finish the sentence. There was a violent burst of orange-yellow from within the darkened woods. It was like a mortar flash, only horizontal, and aimed right at us. It was followed instantly by another, the flame of the second weapon firing lighting up the billowing cloud of exhaust smoke hanging beneath the trees.

Two black streaks each the size and shape of a bowling pin on its side wobbled and skitted their way towards us. One seemed aimed at Sticky's head, the other at mine. In they came. Time seemed to freeze.

Sticky and I dived for cover at the same moment, a Royal Marine and a Light Dragoon doing what had been drilled into us over the years and years of training. As we hit the dirt, the two warheads screamed through the air where we'd just been standing.

The exhaust from the rocket-propelled grenades' sustainer motors enveloped us in a choking fog. The smell was like Guy Fawkes night, only these weren't fireworks. These rockets were designed to shred flesh, and pierce and pulverise armoured vehicles far tougher than our lightly protected Vector.

I swivelled my head to yell a warning. A dirty blue-white trail of smoke lay to either side of the Vector. The RPGs had missed it by inches. *Fucking hell*. The opening salvo of the battle for Adin Zai had been fired, and no guessing who the enemy were targeting. It was us – that 'enemy tank' stuck out on the ridge line.

'RPGs!' I yelled. 'Fucking RPG team in the woodline!'

I turned back to the fight and tugged the angular butt of my SA80 into my shoulder. I jammed my right eye against the smooth metal of my stubby SUSAT sight. Its four-times magnification pulled the enemy position into instant close-up focus, the smoke from the RPG double-tap still hanging in the air beneath the branches.

I placed the diamond-sharp tip of the pointer on the heart of the smoke, and opened fire, pumping round after round into the enemy position. With each squeeze of the trigger a gleaming brass case spewed from the assault rifle's ejector, spinning on to the bone-white rock and dirt beside me. With each I imagined a bullet tearing into an RPG-gunner's skull.

Two more warheads fired out of the woodline. Again, they did the wobble-streak towards us, threading grey smoke across the valley. But this time their aim was a fraction off, the rockets screaming past a few metres above us. From the muzzle flashes I could see that the enemy had moved position a metre or two along the trees. The canny fuckers.

'Watch our tracer!' I yelled to the lads in the Fire Support Group. 'Watch our tracer!'

The FSG's two WMIKs were parked to the side of our wagon. Together they had a pair of 50-cals and two GPMG – 'Gimpy' – machine guns mounted on the open-topped Land Rovers. It should be more than enough firepower to silence those fuckers in the woods. They could follow the glowing threads of our tracer rounds directly on to target.

The crack of mine and Sticky's assault rifles was drowned out by the murderous roar as the Gimpys opened up, followed closely by the awesome thump-thump-thump of the 50-cals. It was deafening, and it felt good.

The RPG teams would keep moving, for their muzzle flashes were a dead giveaway. But the lads on the big machine guns knew their stuff. They were malleting either end of the treeline, trapping the enemy in their positions.

There was a *whump* from beside me, as Sticky loosed off a grenade. Being a rufty-tufty Marine he had the chunkier version of the SA80, with underslung grenade launcher attached. For an instant I caught his eye, to let him know I liked what he was doing.

As I did so, there was a violent kick to my right elbow and an angry high-pitched whine, sand and rocks flying from where the bullet had hit. It left a tiny, smoking scoop in the dirt a couple of inches from my right arm. I rolled to one side, and a second round kicked up the shit where I'd just been lying.

I yelled a warning to Sticky that the enemy had a bead on us. As I did so, I heard a voice screaming from behind me.

'Sticky! Bommer! What the fuck're you doing? Bommer – get in the fucking wagon and get the A-10s on that treeline! NOW! And Sticky, cue up the fucking guns.'

'Bommer' was my nickname from The Light Dragoons. Chris was dead right. Sticky and I had just sacked our stations, getting fire on to the treeline when we should have been doing our proper jobs. I grabbed my rifle and made a mad dash for the rear of the Vector.

'Aye-aye, Johnny Bravo,' I panted, as I dived inside. 'I'm on it.'

Captain Chris Lane, from 19 Regiment Royal Artillery, was commanding our Fire Support Team (FST). He'd earned the nickname 'Johnny Bravo', just as soon as we'd clapped eyes on his shock of blond hair and rippling, chiselled torso. JB had us bang to rights for fucking off from our proper jobs.

As the Joint Terminal Attack Controller (JTAC – pronounced 'Jaytack') attached to B Company, 2 MERCIAN, it was my role to call in the warplanes. I was only to use my personal weapon as a last resort. Trouble was, that was all totally counter-intuitive to normal soldiering.

The natural reaction whenever you were engaged was to put down rounds. To save your life and that of your mates. To kill the enemy. It was what soldiers like Sticky and me had trained to do for years. But as a JTAC I had to force myself to go against all my instincts, and trust the ground troops to defend me.

Since 0400 that morning I'd been working the air power. First, I'd had a pair of A-10 Thunderbolt ground-attack aircraft on station. They'd been ripped by a Harrier and then another pair of A-10s. I had those A-10s overhead right now. One was searching the Green Zone for the enemy, the other checking the compounds ahead of the 2 MERCIAN's line of departure.

There was nothing better to hit that enemy RPG team than those A-10 'tank busters'. The A-10 has a nose-mounted, seven-barrel 30mm Gatling gun that spews out a staggering 3,900 rounds per minute. It provides devastating firepower even against a main battle tank, and would turn those RPG gunners into Taliban purée.

So powerful was the kickback from the cannon, that it had been known to stall the aircraft's giant turbofan jet engines. In theory they could be restarted in mid-air. But I didn't fancy being an A-10 pilot and trying. Either way, the A-10 was fast becoming my aircraft of choice in Afghanistan.

I scrabbled about in the rear of the wagon for the handset of my TACSAT, a UHF ground-to-air radio. The back of the Vector was my domain. JTAC Central. It might look like total chaos, but it was my chaos. My fingers grabbed the TACSAT handset from under the seat, and I jammed it against my ear.

'*Hog Two Two, Widow Seven Nine*, do you copy?'

There was a burst of echoing static in my ear. It was drowned out by a volley of bullets slamming into the compound wall directly behind us, chunks of blasted mud wall hammering off the Vector's armoured sides. I glanced skywards, cursing for the A-10 to respond.

From the TACSAT a black cable snaked out of the Vector's open hatch, connecting to a satellite antenna atop the wagon. From there, the signal beamed skywards to the receiver embedded in the nose cone of the jet. But the TACSAT was a line-of-sight comms system. If the A-10s were out of sight they would miss my call.

'*Widow Seven Nine*, this is *Hog Two Two*, you're loud and clear.'

Yeah! We were on. 'Sitrep: engaging enemy RPG team in north–south woodline two hundred metres due east of our position. Can you see our tracer?'

'Negative. I don't see your tracer,' came the American pilot's calm drawl.

I'd already talked the A-10 pilots around our position. I'd given them the layout of the three platoons below us, and their routes of advance into Adin Zai. Using maps compiled from aerial photos by our GeoCell unit, we'd located the three targets of today's mission – Objectives Silver, Gold and Platinum. The last – Objective Platinum – was a suspected Taliban training school.

I'd briefed the pilots on the weapons systems Intel reckoned the enemy had in there. Apart from the usual – small arms, machine guns and RPGs – there was a B-10 107mm anti-tank gun, a big and nasty bit of kit.

The A-10's one drawback is its speed. Maxed out it only does 420mph, about the same as that achieved by the P-47 Thunderbolt, the Second World War aircraft the A-10 is named after. That left it vulnerable to the kind of fire put up by a 107mm anti-aircraft gun.

I checked the GeoCell map propped opposite me on some ammo crates, searching for the enemy position. For a second it struck me how it would really chafe if those ammo crates got hit by an RPG. I forced the thought to the back of my mind.

I ran my finger along the map east from our position, and found the enemy treeline. I took a slug of flat, lukewarm Sprite from an open can, spat out a couple of dead flies, then spoke into the handset.

'Enemy RPG team in treeline running north–south for one hundred metres from ridge line, then dog-legs south-east for fifty, terminating in a dirt track running west–east along riverside. D'you see it?'

'Visual treeline,' came back the pilot's reply. 'This is what I see: L-shaped wood with smoke plumes at north end, just below demarcation line.'

Result! I glanced at the GPS I had hung from the roof at the front of the wagon, and did a flash of mental arithmetic. 'Enemy coordinates are 67473628. Nearest friendlies – our position two hundred metres west. Readback.'

The pilot read the details back to me.

'Affirm,' I confirmed. 'I want immediate attack with three-zero mike-mike strafing treeline on a south-to-north attacking run.'

I checked the GeoCell map one last time. The tiniest fraction of a mistake – one digit wrong on the coordinates – and I would bring down the strafe on to our position, or worse still that of the lads below. I double-checked the A-10's line of attack: it should keep the 30mm cannon fire well away from us and the 2 MERCIAN lads.

'Tipping in,' the pilot confirmed. 'Requesting clearance.'

It's the JTAC who 'buys the bomb' – always. It's our call to ID and clear a target, to choose the weapons system, and to make sure friendly forces are a safe distance from the strike. Without my final clearance, the A-10 pilot would abort.

I stuck my head out the top of the wagon and searched the sky to the south-east, the direction the A-10 should be attacking from. As I did so, a barrage of rounds started sparking and whining off the armoured roof of the Vector. I needed that bloody strafing run, and I needed it now. But I couldn't see the A-10 anywhere.

I caught the distant flash of sunlight on metal. It was an aircraft, but it wasn't where it should be. It was coming out of the rising sun directly to the east of us. On that line he'd be hitting the treeline cross-wise, which was no fucking use to anyone. Worse still, we were directly in his line of fire. Those 30mm cannon rounds would make mincemeat out of the Vector.

I yanked the handset up on its lead. 'Abort! Abort!' I yelled. 'Abort! Fuck off and attack from south to north! Attack line as instructed!'

'Roger, aborting,' the pilot confirmed. 'South-to-north attack run. Banking around now.'

I watched the pilot pull out of his dive, and roll the aircraft into a tight left-hand turn. Thank fuck for that. The squat ugly form hadn't been nicknamed 'the Warthog' for nothing. It was neither graceful nor pretty, but as a ground attack aircraft it has no equal.

'Repositioning,' the pilot confirmed. He popped the jet, bringing it up in a screaming climb. 'Visual with enemy pax in the woods,' the pilot continued. 'The treeline is twenty metres across, and I can see armed figures running around in there.'

From the front of the Vector I could hear Chris briefing the OC, Major 'Butsy' Butt, on all that I was doing. The OC was a blinding commander, and he had total trust in the air power. Previous contacts had proven what a battle winner it was.

Still, he was down in the bush of the Green Zone, having gone firm on the start line, and he had to be wondering what the fuck was going on. Before his men had even begun their advance we'd started world war three up here, and there were rounds and RPGs and jets screaming through the air.

It was Chris's job to keep the OC informed of all that I was doing. He was monitoring my frequency, which was reserved solely for JTAC-to-air comms, and relaying all to the OC, which left me clear and focused to call in the warplanes.

The A-10 reached the top of its climb and keeled over, coming nose-down on to the enemy position. The seven-barrel cannon fires wherever the jet is pointing, so the pilot has to dive directly on to target.

'Tipping in,' came the pilot's drawl. 'Requesting clearance.'

At that moment a third pair of RPGs came howling towards the wagon. I ducked, my head and shoulders buffeted by the powerful shockwave as the rockets howled past the Vector's open turrets. They were fag-paper close to us.

'No change friendlies!' I screamed into the handset. 'You're clear hot!'

'In hot,' came the reply. 'Engaging.'

I was oblivious to the enemy gunfire now. I had to see the attack go in. It was crucial for the JTAC to confirm the success of any airstrike. The pilot might report a good hit, but the aircrew didn't always see everything. To wrongly report a target eliminated and allow your lads to advance could cost many good men's lives.

The A-10 seemed almost to stall in mid-air, as the Gatling gun opened fire. There was a long, thunderous 'brrrrrrrrrrrrrrrrrrrr', as the roar of the seven-barrel cannon echoed around and around the valley. It sounded like one of those automatic machines that counts tenners at the bank, only magnified a thousand times over.

The thick, stubby barrel of the Gatling gun was clearly visible spitting fire. For every second the pilot kept his finger on the trigger mechanism, sixty-five 30mm cannon rounds tore into target.

Shredded branches and jagged chunks of tree trunk spewed out of the woodline, along with god-only-knew what. It was like Farmer Giles was getting to work in there with some massive, ghostly hedge-cutter. By the time the pilot had bottomed out the Warthog's dive, he'd raked the entire woodland from end to end.

'*Hog Two Two*, that was a class strafe,' I radioed. 'Fucking class.'

I checked with his wing aircraft, *Hog One One*, but he was busy at the refuelling tanker. I had the pair of them for two hours on 'yo-yo', meaning they were flying a relay, taking on avgas at a refuelling aircraft so as to return to the battlefield.

I radioed the first pilot. '*Hog Two Two*, I want immediate re-attack, the woodline spraying on same line of attack.'

'Roger that. Banking around now. Tipping in.'

The pilot put his aircraft into a tight climb, the jet engines screaming away like a pair of overworked giant hairdryers. As he reached the apex and threw the aircraft into a steep dive, I cleared him in.

The stubby muzzle on the A-10's nose spat fire. The pilot did a second, even longer and more devastating strafe.

'Brrrrrrrrrrrrrrrrrrrrrrrrrrrrrrrrrrrrrrrrrrrrrrrrrrrrrrrrrrrrrrrrrrrzztt.'

As the last echoes of the cannon fire faded away, it suddenly all went very quiet. Our own 50-cals and Gimpys had ceased firing. And for the first time since that opening RPG volley, we weren't being pounded by bullets and warheads any more.

For now at least, the valley of Adin Zai had fallen into an eerie silence.

TWO
CARNAGE

It was barely 0700 and the 2 MERCIAN lads had yet to set foot into bandit country. But my heart was beating as fast as an A-10's Gatling gun, my pulse booming in my ears. The boiling Afghan heat was yet to hit us, but I was already sweating like a pig.

I glanced at Sticky beside me in the wagon's turret, then Chris below up front. Everyone's eyes were like saucers.

'Fookin' hell,' I muttered, doing my best Redcar accent. 'Fookin' fookin' fookin' 'ell.'

I'm from the north-east of England, and everyone mistakes me for a Geordie. I was forever playing up to it with the three southerners who made up our FST.

Sticky smiled. 'You took your time to hit 'em, mate.'

'Result,' Chris added.

'Sorted,' I confirmed.

'You have to, mate?' Throp grunted, from where his bulk was half-hidden, hunched over the Vector's wheel. 'I was enjoying the scrap.'

It was typical Throp. Like Chris, Lance Bombardier Martin Hemmingfield was a Royal Artillery lad. He was a six-foot-two hunk of muscle and bone, with the breaking strain of a KitKat. He was also completely fearless. You could put him in front of ten meatheads in a street brawl, and stand back as he took them all on.

I loved having Throp on the team, but I never got to the bottom of why everyone called him 'Throp'. He claimed it was something about 'Hemmingfield' sounding like Hetty Wainthropp off *Hetty*

Wainthropp Investigates, some BBC series about a pensioner solving minor crimes. Wainthropp had been shortened to Throp, and that became his name. Or something. Anyhow, who would want to argue with Throp about why he was called Throp?

There was a noisy upsurge in radio traffic. Reports were coming in of the enemy chatter thick and fast. There were increasingly desperate calls for 'Hamid' to check in with the Taliban commander. Each call was met with an echoing void of static. If Hamid was still alive he certainly wasn't answering.

With the enemy guns having fallen silent, Major Butt ordered his men up and into the advance. Twenty minutes later the three platoons had pushed five hundred metres into the Green Zone on foot, with not a shot having been fired. If it carried on like this we'd be back at base in time to get the kettle on for breakfast.

The two WMIKs left us and moved up to what remained of the RPG-gunners' position. They radioed in reports of blood-spattered undergrowth, but no bodies. The enemy were good at collecting their dead. In an effort to show that we came not to kill but to fight when attacked, we'd leave them to do so unhindered. It seemed wrong not to.

The A-10s were ripped by a lone Dutch F-16, call sign *Rammit Six Two*. I was halfway through giving the pilot an Area of Operations (AO) update, when it all went bloody bananas. The Company HQ had been ambushed at close quarters. Major Butt and his men were deep in the Green Zone on foot, and getting smashed in there.

Butsy radioed in that four enemy had been killed, but that fighters kept coming. From the turret of the wagon I couldn't make out a bloody thing. I could hear the crack and thump of battle, and see the odd flash of desert camo as our lads tried to get into position to fight. But I couldn't see how I could drop any bombs or strafe. The contact was beyond danger-close. Plus I couldn't ID the enemy to hit them.

It was a classic Taliban ambush. Their favourite tactic was to let a force advance, whilst outflanking them. Then they'd capture or kill the lot. There was no way I was going to let that happen to our lads. Most of the 2 MERCIANS were late teenagers or in their early twenties. They were fresh-faced and only a few weeks out of the UK.

Many had never left home before, but they were in the British Army and they were under orders. They'd come thousands of miles to soldier on behalf of the Afghan people and their fellow countrymen back in the UK. They were brave and tough, despite their youth, and they'd fight to the very last for their brothers in arms.

They weren't my regiment, but that didn't matter. JTAC-ing is a highly specialist role. There's never more than a couple of hundred qualified JTACs in the entire British Army. In theatre we'd get embedded with whatever unit needs us. Pretty much from the start of their tour I'd been with B Company, 2 MERCIAN. I'd bonded with those young lads, just as I'd bonded with my FST. *I was their JTAC.* And as far as I was concerned they were my boys, and I felt responsible for every last one of them.

Those 2 MERCIAN lads were some of the best infantrymen around – England's finest. I wasn't about to let them get injured, or fall into the hands of the enemy if I could help it. A few days back we'd been told a story about some elite French commandos on some hush-hush mission. Somehow, eleven of them had got themselves captured by Taliban, or more likely Al Qaeda elements. None of the French captives had made it out of enemy captivity alive, and I shudder to think what had happened to them before they were killed.

We knew what the enemy were capable of, and now we had B Company's HQ element about to be overrun in Adin Zai. In a way I wasn't surprised. I knew what Major Butt was like. You couldn't wish for a better OC. Butsy was always taking himself and his HQ element into the heart of the action.

Major Butt led from the very front. *Always.* The OC was always to be found in the thick of it, with just the four men of his HQ element as security. Butsy was a legend, and no way was the OC getting captured and tortured or killed on our watch. The question was, what the fuck were me and *Rammit Six Two* going to do about it?

Butsy's signaller was giving a running commentary on the firefight over the net. From the open turret of the wagon the noise of battle was deafening. In part the signaller was hoping that our four-man FST would hear him, and find a way to get them out of the shit. No one else had a hope of doing so, that was for certain.

Trouble was, a lot of the firepower at our disposal was a pretty blunt instrument. We had a battery of 105mm howitzers that Sticky, Throp and Chris could call on to target, plus we had the mortar teams. But with the HQ element surrounded, we risked shelling our own men.

Air power was the precision killing machine. But not even the state-of-the-art F-16 Fighting Falcon I had above me could do much right now. Our lads were tens of metres away from an invisible enemy. No JTAC would risk calling in ordnance in such danger-close conditions.

An idea came to me. It was something I'd learned about in JTAC school, back in the UK. The JTAC course is far from easy, especially for someone like me. I'd left school at sixteen, and I'm the first to admit I'm no Einstein. The course has a high dropout rate, and I was hardly your ideal candidate. But my CO at The Light Dragoons had backed me all the way.

At JTAC school I'd had to concentrate real hard, plus do my homework like a good lad. I'd forced myself to come forward in class and ask questions, and not to care if anyone thought that I was a thick fucker. It was the kind of thing that I should have done at high school, had I not been too busy trying to get into the army instead.

A couple of weeks after getting JTAC combat-ready, I'd deployed to the war in Afghanistan – so I'd not had a chance to try in practice what I'd got in mind. But right now it was the best I could think of. I got Sticky to radio Butsy and ask for the exact, ten-figure grid of his position. Meanwhile, I got on the air to the F-16.

'*Rammit Six Two, Widow Seven Nine*. Sitrep: our HQ element is surrounded and engaged at close quarters in the Green Zone. This is the HQ element's grid.' I took the scrap of paper Sticky thrust at me and read the numbers. '46673896. Repeat: 46673896. Readback.'

The F-16 pilot read the details back to me.

'Affirm. I want you to fly an immediate show of force over that grid. I want you to come in lower than a snake's belly and achieve sonic boom right on top of the HQ position.'

'Roger. Show of force at seventy-five feet – lower than a snake's belly.' I heard the Dutch pilot chuckle. 'Tipping in.'

It was rock-hard for a pilot to achieve sonic boom on demand. Only the best of the best could do it. He'd have to put the jet into a steep dive and pull up violently, the collision of the jet engine's thrust and the air creating a massive boom.

I learnt about it at JTAC school, and the sound was like a couple of thousand-pound bombs going off. For good measure I asked the pilot to pop flares, so it looked like he'd fired off a volley of missiles.

'Radio the OC,' I yelled at Sticky, who was on my shoulder in the Vector's turret. 'Brief him on what I'm doing.'

In one ear I listened to Sticky warning Butsy that the F-16 was preparing to fly a low-level show of force, no ordnance. In the other, I had the F-16 pilot talking me through his actions, as he came around on to target.

As the jet screamed in the enemy would get their heads well down. Or at least that was the plan. At that moment Butsy and the HQ element were going to bug out. It was the best I could think of to get the lads out of the shit.

The pilot banked his aircraft in a screaming turn, the wingtips trailing jets of white. Then the F-16 was tearing in like a thing possessed. The arrow-like streak of the aircraft seemed to touch the very tree tops as it reached the lowest point of its dive. As it pulled up sharply it popped flares, leaving a shower of glowing sparks like a comet it its wake.

I covered my ears and cringed in the turret, waiting for the sonic boom. But none came. Well, I had been asking the earth of the pilot, and it was still a wicked show of force. The ear-splitting scream of the jet engines rattled and shook the Vector, as each flare floated in the air like a tiny, blinding white fire.

I could sense the drop in the intensity of the gunfire, before the dying roar of the jet engines allowed me to hear it. From below me Chris was relaying the radio chat, as I kept my net open with the pilot. If the show of force had failed, I was going to have to dream up something else with that pilot.

'Major Butt reports contact has gone quiet,' Chris relayed. 'Six enemy killed … No more incoming fire … Enemy has withdrawn … Extracting from their positions …'

Phew! We were out of the shit. Time to crack on with the mission.

The F-16 was low on fuel, and it was ripped by a singleton F-18 Hornet, call sign *Uproar Two One*. The platoons below me were on the advance again. As I talked the pilot around the battlefield, the valley had fallen ominously silent.

I guided the pilot around our position and that of the troops below, plus I gave him a heads-up on their line of advance. Sticky and I glanced across the Green Zone to where the F-18 was circling. For a brief moment we enjoyed the relative quiet that had settled over the battlefield. The calm before the storm.

I reached down and made a grab for my piss bottle through the open turret. It was a golden rule of JTAC-ing: *never miss an opportunity to urinate*. Nothing was allowed to get in the way of the JTAC-to-air process. It was too far to reach the bottle, so I asked Sticky to pass it me.

'Thimble bladder strikes again,' I quipped.

Up through the turret came a nasty-looking 1.5-litre plastic water bottle, with the top cut off. Over the past few weeks I'd learned that trying to pee through the neck of the bottle was a bad idea. So Sticky had taken on the role of chief bottle-cutter.

I did my business and passed it back to Sticky, who threw the contents out of the wagon's back door. The last thing we needed was someone knocking it over before it was emptied. It was smelly enough in the wagon as it was.

Off to the east there was a faint boom. I jerked my head up and scanned the horizon. About 2.5 clicks due east was a tiny plume of smoke. I glanced at Sticky. He shrugged. I guessed he was right. Whatever it was, it was far distant from the battlefield and hardly a threat.

I got on the net and briefed the F-18 pilot, just in case. As I spoke into the TACSAT I became aware of a faint whistling. For a couple of seconds I mistook it for comms interference, and then the whistle became a piercing scream. An instant later the mortar round slammed into the ridge line.

It impacted seventy-five metres east of the Vector, the blast showering us in rocks and sand. There was another distant boom and a puff of smoke, and a second mortar came howling down. It landed sixty-five metres to the west of the wagon, blasting Sticky and me in shit, and hammering the wagon's steel sides.

The enemy mortar boys were good. They had to have a dicker somewhere spotting where their rounds were falling. He'd have a mobile phone and be calling in adjustments to their fire. Now they had us bracketed, with a mortar dropped on either side of us. Top joy that was.

Shell number three had our names written all over it. We dived into the wagon, as Throp revved the engine. He floored the accelerator, and the Vector growled and shook itself into motion, the six wheels crunching and spinning backwards away from the drop-off.

Chris was on the radio, warning the OC that we were under mortar fire, and moving south-west along the ridge. It wouldn't make one hell of a lot of difference to those mortar boys. We'd still be sticking out like a bullet magnet, and well within their range. Either we took out that dicker or the mortar team, or they were going to get us.

There was no time to worry about it. As the third mortar ploughed into the dust-dry earth where the wagon had just been standing, all hell broke lose below. 3 Platoon – call sign *Arsenic Three Zero* – had been ambushed from sixty metres. It was 0800, and the lads had yet to hit any of the objectives, and they were getting smashed again.

The F-18 Hornet above me packed an array of top-notch weaponry, including a 20mm cannon and Maverick air-to-surface missiles, plus precision-guided JDAM and Paveway 'smart' bombs. But the danger-close distance to which any of those could be deployed was measured in hundreds of metres – not the sixty between the lads of 3 Platoon and the enemy.

I put a call through to the pilot. '*Uproar Two One*, *Widow Seven Nine*. Sitrep: I have a forward platoon engaged at close quarters. This is the platoon's grid: 93467235. Readback.'

The pilot confirmed the grid.

'Affirm. I want you to fly a show of force over the grid. We're under mortar fire, so how low can you bring your jet without getting hit?'

'How low d'ya want me?' came back the American pilot's reply.

'As low as you can get. I want a low-level pass with flares and sonic boom.'

'Roger that.'

'What about the mortars?' I queried.

—

'*Widow Seven Nine*, I don't give a fuck. It's a big sky small target. Tippin' in.'

I turned to Sticky and grinned. 'Fucking top bloke, or what!'

The F-18 came screaming in popping flares as it went, and if you'd thrown a rock you could have hit it, it was so low. Right over the grid the pilot pulled up violently, the air in the jet's wake like a tortured steam cloud. A massive deep BOOM! thundered the length and breadth of the Green Zone, and I could feel the Vector beneath me shiver with the shockwave.

Now *that* was a sonic boom.

Achieving sonic boom uses up shedloads of fuel. The F-18 pilot warned me he had to head directly for the refuelling tanker. He got ripped with a Dutch F-16, call sign *Rammit Six Three*. The Dutch jet was inbound into my Restricted Operating Zone (ROZ) four minutes out. He'd enter my ROZ just as the F-18 was leaving it.

A ROZ is a block of airspace above a battlefield that is the exclusive domain of a JTAC. No other JTAC is allowed to operate in that ROZ, and no aircraft other than those controlled by that JTAC are permitted to fly in it. This allows for deconfliction between air assets, and prevents one aircraft flying into another, or getting hit by 'friendly' bombs.

I briefed *Rammit Six Three* on the battle as he was inbound. 3 Platoon were still under withering fire. I asked the pilot to fly an immediate low-level pass over their grid as soon as he was with us, firing flares. But I sensed the enemy were getting wise to these shows of force.

I was getting well pissed off. I knew where the enemy were, yet I couldn't kill them, for they were too close to our lads. What I needed was Apache. I put a call through to Widow Tactical Operations Centre (TOC), the central command element for all air operations.

'*Widow TOC, Widow Seven Nine*,' I rasped.

'*Widow Seven Nine, Widow TOC.*'

'Sitrep: as at now – TIC.' TIC stands for Troops in Contact, the minimum requirement to call out Apache gunships. 'I've got a platoon in close-quarters combat in the Green Zone.'

As I spoke into the TACSAT there was a deafening explosion right on my left shoulder. Yet another mortar had slammed into the ridge line barely metres from us. It hadn't taken that mortar team long to retarget our wagon.

'*Widow TOC*, wait out!' I yelled.

I grabbed the edge of the turret and clung on tight, as Throp gunned the Vector through the cloud of blasted smoke and sand, doing his hide-and-seek routine with the enemy mortar team.

He wrenched the wagon to one side, crunching over boulders and the broken masonry of a half-demolished wall. He pulled up in a flattened and deserted compound, gaining us a little cover.

On the net I could hear the leader of 3 Platoon calling for airstrikes. The enemy were pressing in on his position. I fucking needed those Apache yesterday.

'*Widow TOC, Widow Seven Nine* – requesting immediate CCA!' I yelled into the handset. 'Repeat: immediate CCA.'

CCA was the call to launch Apache. '*Widow Seven Nine, Widow TOC*: affirm: CCA, fifteen minutes to launch. Repeat: fifteen minutes to launch.'

The AH-64 Apache gunships would be in the air in fifteen minutes, which would get them over Adin Zai in twenty. That's how long the lads of 3 Platoon had to hang on before we had some surgical air power above us. I got Sticky to tell them what was what, and they reported back five enemy fighters killed. But still they were pinned down and taking murderous fire.

The Apache helicopter gunships can engage targets closer than any other air asset. They can sit at altitude eyeing their sniper optics and cuing up their 30mm cannon, and doing danger-close engagements. With jets it was all about positioning attack runs, and with their speed of approach they were far less accurate. That's what I'd learned at JTAC school, and that's how it had proven over the last few weeks in Helmand. But it was a lesson the enemy also seemed to have learned well.

I got allocated two Apaches, call signs *Ugly Five Zero* and *Ugly Five One*. I got the F-16 banked up high, to allow the gunships in to do their work. But as soon as the squat black predatory shapes of the Apaches tipped up over the battlefield, the contact died down.

It all went deadly silent. I sat in the Vector's turret hardly believing what was happening. *Nothing*. There wasn't a single round being loosed off below, or a mortar being lobbed at us lot on the high ground. It was as if I'd called out the Ugly call signs on a lie.

The enemy seemed to have vanished off the face of the earth.

THREE
GRAND THEFT AUTO

I split the flight. I got *Ugly Five Two* scanning the compounds to the front of 3 Platoon, whilst *Ugly Five Zero* got overhead the enemy mortar plate. Not a soul was moving on the ground, and there wasn't a thing to be seen in either location. Nothing. *Nada*. Zilch.

At least it gave 3 Platoon time to re-bomb their mags and get some water down their necks. It was late morning under a burning Afghan sun, and the lads in the Green Zone had to be sweating their cocks off. It was like an oven in the Vector. It had zero air-conditioning, which was another reason we kept the turrets open.

By rights we shouldn't even have been using an armoured truck like the Vector. On arrival in Camp Bastion our FST had been issued with a WMIK – an open-topped vehicle bristling with machine guns. It sure looked the business, but after a handful of missions we'd realised how utterly useless it was for our tasking.

The WMIK is a three-seater vehicle: two at the front and one at the rear on the 50-cal. Someone seemed to have overlooked the fact that we were four in our FST. Smart. Whenever we went out on ops we had to leave one behind. Plus using an open-topped Land Rover in the burning heat and dust killed the kind of kit we were using. The tools of the trade are all kinds of sensitive electronic equipment: radios, satellite comms, laser target designators, computers and handheld navigation devices. The kit was taking a right hammering in the WMIK, and it wouldn't last the duration of the tour.

A month into our deployment Sticky, Throp and I had been rotated back through Camp Bastion. We'd been promised a Vector at the very least, and we had our hearts set on getting one. But when we got there we were told that 'our' vehicle had been allocated to another unit.

So we went looking for one. We found a Vector complete with its work ticket – a green leather wallet containing all its roadworthy certs – and with the keys in the ignition. It had a full tank of fuel, so we jumped aboard and joined a convoy heading back to base, at FOB Price.

Admittedly, we'd 'borrowed' that Vector, but no one seemed to mind. And ever since then we'd been driving around in 'our' Vector. It was about the minimum that the four of us plus all our kit could get away with.

And I was mighty glad of it now, stuck up here on the ridge line overlooking Adin Zai. I'd lost count of the number of times we'd been bracketed by the enemy mortar team, shrapnel and rocks slamming into the steel sides of the wagon.

By now we were also getting targeted by a 107mm rocket launcher. A direct hit from a 107mm warhead would be terminally lethal. Plus small arms rounds kept pinging off the armoured skin of the beast. The Vector mightn't be bomb-proof, but it sure as hell was doing its job up here on the high ground.

By 1300 there was still not the slightest sign of the enemy. The Apache gunships were low on fuel, and they left to return to Bastion without a shot being fired. We were yet to have a single injury amongst our lads, which was unbelievable. We'd been lucky as fuck, and having the Apaches overhead had bought us some precious time.

I told Widow TOC that I needed air cover, as we were TIC-imminent. I got allocated an American F-16, *Wicked Four One*, which was five minutes out. As I awaited the jet's arrival, I got down from the wagon. Sticky, Chris and I strolled around to the front of the Vector, wondering how it could all have gone so silent.

'It's fucking spooky,' Sticky remarked, as he gazed over the Green Zone. 'Where've they gone? They've just disappeared.'

'Aye – that's Apache for you.' I dug in the pocket of my combats and pulled out a packet of ciggies. 'I got to feed me habit.' I waved the packet around. 'Anyone?'

Chris was a fitness fanatic, and too much into being body beautiful to smoke a tab. Throp would share the odd Lambert & Butler moment with me, but there was no getting him out of his seat at the Vector's wheel. It was fair enough: his rapid manoeuvring had saved us from getting splatted more than once that morning.

I slumped down on the dirt, leaning my back against the wagon's knobbly tires. God, was I knackered. I was hanging out of me hoop. We'd been on the go for forty hours, and for two nights I hadn't slept. It was only the adrenaline that was keeping me wired.

It was the morning of 16 May – two full days ago – when Butsy had first briefed us on the Adin Zai mission. Chris insisted on the whole of the FST being present during briefings, and each of us worked directly to the OC's orders. Chris wanted every one of us to hear what the gaffer had to say, in case one or another of us was taken out during the coming battle.

The key aim of the Adin Zai mission was to take Objective Platinum – the Taliban training school. That was our limit of exploitation, and we would push no further east. We were targeting an enemy stronghold in the heart of 'their territory' – the Green Zone – and I had priority as a JTAC throughout Helmand in terms of air missions.

At 2300 we'd pushed out of the British base at FOB Price in a convoy of Vikings, WMIKs and Snatches, plus our borrowed Vector. I had an intelligence asset flying over the convoy, call sign *Dragon Zero Two*. No sooner had the gates of FOB Price clanged

—

shut, than the aircraft started picking up some interesting snippets of enemy chatter.

'The enemy tanks have left their base!' the Taliban were yelling to each other. They called all our vehicles 'tanks', no matter what they were. 'They're heading to Adin Zai. Fight them to the death, brothers! *Allahu akbar!*'

So much for the secrecy of our mission. The enemy seemed to know what we were doing almost before we did. I had air controls all that night as we crept through the open desert. By the time we reached the high ground near Adin Zia, the air cover was reporting women and children fleeing the village to the east. It was a classic combat indicator.

By mid-morning, Adin Zai village was totally deserted, apart from groups of males of fighting age. But this wasn't going to be any old gunfight at the OK Corral. Under the rules of engagement we had to PID (positively identify) enemy fighters before killing them, and ideally once they'd started shooting at us to prove 'hostile intent'.

We'd lain up in the open desert during the day, our orders being to launch the attack at first light the following morning. I'd had air platforms stacked up above me, flying recces over the mission objectives.

'We await the tanks that are parked in the desert!' the Taliban commanders were calling to each other. 'Hold firm in your posi-tions, brothers, until they move to attack us!'

I'd had more air that night, and a result I hadn't slept a wink. Now, as I sucked nicotine into my greedy lungs, I realised how totally and utterly chinstrapped I was. I lay on the dirt longing to close my eyes and get just a few minutes' kip.

I jerked awake to the sound of a 107mm Chinese rocket screaming over the top of us. I'd dropped off for a second or two, my chin nodding on to my chest. The warhead ploughed into the desert some seventy metres beyond us, throwing out a deafening blast and a cloud of dirt and smoke. What a fucking rude awakening that was.

I heard Sticky's laugh. 'Yeah, and no guessing: the next will be short.'

Each 107mm rocket was about the size of a man's leg. It took two to slide the twenty-kilo warhead into the launch tube. The enemy were using a man-portable tripod launcher, hence the time between each rocket being loaded, re-aimed and fired. So now we had a mortar unit, plus a 107mm launcher team to find and smash from the air.

Not a minute after that first rocket had been fired, a second came screaming down on us. It smacked into the rock of the ridge line less than twenty metres below. It threw up an angry mushroom cloud of black smoke high above our heads.

Sticky let out a crazed cackle. 'Didn't I bloody say so!'

'Right, in the wagon!' Chris yelled. 'Let's get moving.'

I took a last drag on my tab, flicked the butt away and levered myself to my feet. As I turned to clamber aboard the Vector there was a howling, screaming inrush, like a bloody great big dragon was about to breathe fire down our necks. An instant later the ground shook with a sickening, thudding impact right at my very feet.

That third 107mm ploughed into the dirt three metres from where Chris, Sticky and I were standing. *This is it*, I thought. *They got us. We're fucking dead*. I tensed for the explosion, fully expecting to see Sticky and Chris's brains splattered all over the side of the Vector, an instant before mine joined them.

Instead, a choking cloud of dust and sand engulfed us. I felt like shit, but I sure as hell wasn't dead. Gradually, the cloud cleared. It revealed a small crater more like a rabbit hole right in the shadow of the Vector. I stared at it, barely daring to breathe. I swallowed hard.

That 107mm rocket had burrowed a hole in the dirt at our very feet, but it hadn't exploded. *It was a dud. It was a fucking dud.*

I heaved myself into the wagon and grabbed another tab with a hand that was visibly shaking. I sparked it up and clamped it between my teeth to stop the shakes from showing. As I dragged in the smoke, all I kept thinking was this: *What were the chances of that happening? What were the chances?*

It was a direct hit, *and it was a fucking dud*. It was the first – and, as it would prove, the last – dud 107mm of our entire Afghan tour. Whoever says that no one is looking after us? Someone was up there, that was for sure. We had an angel on our shoulders.

I glanced furtively at Sticky, Throp and Chris. No one was saying a word. What was there to say, apart from something crass like: *Aye, well, that was a close one. Best to crack on*. But I knew what they were thinking, 'cause I was thinking it too: *it's time to get the fuck off this ridge line*. No one wanted to be the one to say it, to voice the unthinkable. If we left the high ground, we were as good as abandoning the 2 MERCIAN lads in the midst of the battle of their lives.

'If we get off the bloody ridge we're next to bloody useless,' I muttered, into the hands cupped around my fag. 'I'd rather have a 107mm rammed up me grinner than do that.'

There was a ripple of nervous laughter. 'Grinner' is northern slang for backside. The lads were starting to learn a little of my lingo as time went by.

'Best pray it's another dud, if you've got it up your ass,' said Sticky. 'I still ain't pulling it out though.'

And that was it – nothing more was said about our closest ever encounter with death. No one wanted to stay on this cursed ridge line, that was for certain. But we sure as hell weren't leaving until the job was done.

Lance Bombardier Ben 'Sticky' Stickland was always larking about, no matter what shit we were in. Over the past few weeks he'd become like my mucker. I couldn't help but love him. Sticky hailed from 29 Commando, and being a Commando gunner he was a top soldier. He was fit as a robber's dog, and was on the scale of an ultra-marathon runner. He had a totally stupid sense of humour, and was the funniest kind of killer you ever could meet. He and I were always messing, and he was almost as much of a practical joker as me.

But it wasn't this that endeared me most to Sticky. Over the past few weeks he'd become like my shadow. More than anyone, I reckoned Sticky really got it. He could sense the unique beauty of conducting the air war, and he thrilled to the awesome power wielded by the JTAC.

The air war could turn a battle within seconds, and Sticky knew it. No other soldier on the battlefield could bring hundreds of million of dollars' worth of state-of-the-art war machines to bear, with unrivalled power to crush and smash an attacking enemy. He loved JTAC-ing almost as much as I did.

Throp wrestled the wagon across a patch of rutted terrain, trying to choose a spot on high ground where we hadn't already been targeted. A dog should never return to its own vomit, but we didn't have much choice up here. I got on the TACSAT to the F-16 orbiting overhead.

'*Wicked Four One, Widow Seven Nine.* Sitrep: we're getting targeted by mortars plus 107mm rockets. Push up to twenty-five thousand feet, and search around our forward platoons, now eight hundred metres into the Green Zone.'

'Roger that,' came the F-16 pilot's reply. 'Climbing to twenty-five thousand.'

I wanted the F-16 high enough so no one could hear him, but low enough to spy on the enemy. Via the high-resolution optics located in the aircraft's nose cone, the pilot could maintain eyes on the ground even from that altitude. I wanted the enemy to think our air cover had gone, so we could lure them out and smash them.

As the noise of the F-16 faded away to nothing, I levered myself up into the Vector's turret. No sooner had it gone quiet, than there was another distant boom. The mortar was back in action.

But this time, the flash of the exploding round was down in the Green Zone, somewhere around the forward line of our troops. At about the same moment there were a series of sharp cracks of small arms fire and the crump of exploding RPGs.

'*Arsenic Three Zero*'s in contact,' Sticky yelled across at me. 'Small arms, RPGs and mortars. They're pinned down and going nowhere.'

I grabbed Sticky's radio handset. He carried a specialist bit of comms kit that could talk to the 105mm field guns back at FOB Price, Camp Bastion, the air power and most stations in between. At present he had it tuned in to the company net.

'*Arsenic Three Zero, Widow Seven Nine*,' I yelled. 'Give me a grid of your position and talk me on to the contact point.'

'Roger. Grid coming up. Stand by.'

As the platoon commander spoke I could hear the whipcrack of rounds in the background. I could just imagine what that poor bastard was going through. He was in the midst of a shit fight, with thick vegetation all around him and terrified of losing some of his lads. And now he had some arsehole of a JTAC telling him to get out his map and compass and try to work out where the hell they were positioned.

I waited for his reply, with one ear scanning the TACSAT for any comms from the F-16. At the same time I got Sticky to grab the GeoCell map and spread it out on the roof of the wagon. Fuck any rounds that were coming our way. It was time to nail these bastards.

'*Widow Seven Nine, Arsenic Three Zero*.' The platoon commander sounded breathless, like he'd been running. 'Friendly grid: 986745. Repeat: 986745. I've got eyes on the enemy firing point. They're one-seven-five metres due east of us.'

Yeah! Get in! It was still danger-close, but 175 metres was good enough. It was time to smash 'em. I bent over the map, trying to convert the six-figure grid the platoon commander had given me to an eight-figure grid, the minimum the F-16 pilot would need.

'Describe the enemy position,' I yelled into Sticky's handset.

'Treeline one-seven-five metres to the east,' the platoon commander yelled back. 'Running north-west to south-east. There's a kink at the southern end like the handle of a walking stick.'

'Roger. Out.' I passed the handset to Sticky, confident that he would do what was needed. He'd warn 3 Platoon when the bombs were coming in, and Chris would brief the OC.

I pressed the TACSAT to my ear, and dialled up the F-16. I gave him a sitrep, passed him the eight-figure grid of the friendlies, described the enemy position, and told him to get visual with that treeline.

'Comin' down for a closer look,' came the pilot's reply. 'Zooming in my optics to your coordinates. Right, I'm visual with the friendlies.' There was a moment's pause. 'Now visual with muzzle flashes coming out of a dog-legged treeline, one-seventy metres east of there. Visual with heat spots in that treeline. Six pax at least.'

'Stand by to attack,' I replied.

'Jet visual with enemy pax in the treeline!' I yelled to Chris. 'Danger-close one-seventy metres to 3 Platoon. I need OC's clearance.'

With a danger-close mission I needed top-level clearance. As Chris dialled up the OC, I could hear Sticky briefing 3 Platoon on what was happening. He didn't know what bombs I'd be using, and in truth neither did I. I was running through the ordnance package of an F-16 in my head, and trying to work out what was best at 170 metres.

'OC says to hit 'em!' Chris yelled up at me.

We were on! '*Wicked Four One, Widow Seven Nine.* I want you to hit those heat spots with a GBU-38. Repeat: GBU-38. I want you coming in on an attack run ...'

I gave the pilot a bearing that should throw the blast away from 3 Platoon. A GBU-38 is a five-hundred-pound JDAM (Joint Direct Attack Munition). It's a standard Boeing Mk-82 'dumb' bomb, turned 'smart' by having the JDAM precision guidance system strapped on to it.

Standing on end, a GBU-38 is about the same height as Throp, and twice as nasty. It's not the heaviest bit of kit the F-16 carries: the thousand-pound JDAM makes double the noise and blast. But it was about as big a bang as I felt I could risk at 170 metres danger-close to our lads.

'Visual with six enemy pax in the treeline firing RPGs and small arms,' the F-16 pilot radioed. A pause. 'I'm sixty seconds out.'

'Sticky, give 3 Platoon the sixty-second call!' I yelled. 'And check they've not changed position.'

It was bloody hectic now. I had less than a minute to do a visual check of the F-16's attack run, check 3 Platoon hadn't moved, and make the call to clear the airstrike in or abort it. I glanced to the north-east, and bang on cue there was the knife-sharp wedge of the F-16 arrowing out of the burning blue of the Afghan sky.

'Call for clearance,' intoned the pilot.

I glanced at Sticky. He gave me a smile and a thumbs-up.

'No change friendlies,' I told the pilot. 'You're clear hot.'

'In hot,' the pilot confirmed.

There was a tense silence in my handset, as the pilot powered in towards the release point.

Then: 'Stores.'

Sticky radioed the warning to 3 Platoon: 'Bombs away!'

I saw the jet pull up over the release point, and then it was streaking past right in front of our noses. I didn't see the bomb fall, but the flash of the impact was like an ammo dump blowing in a Second World War movie. An instant later the awesome roar of the explosion swept over us, followed by the air-rush of the shockwave.

—

'Fuckin' hell!' I yelled. '*Get in!*' I turned to Sticky. 'Get a sitrep from 3 Platoon.'

The lads knew the bomb was going in, so they'd be on their belt buckles hard in cover. And the airstrike had looked to be bang on target. But the splinter distance – the safe range for friendly forces – of a GBU-38 is 275 metres, and that's with the good guys in proper cover. I wanted to make totally sure the 3 Platoon lads were still alive.

'*Wicked Four One*, BDA,' I radioed the pilot. 'Repeat: BDA.'

I was asking the F-16 pilot for a Battle Damage Assessment (BDA). I didn't really need one, for the contact had died down to nothing. But with his sniper optics he was sure to see more than any of us lot.

'Ground troops are all A-OK,' Sticky reported back to me. 'The impact point was right on top of the enemy. Platoon commander was visual with three enemy with RPGs as the bomb hit 'em.'

'*Widow Seven Nine*, BDA,' the F-16 pilot cut in. 'The only thing left is a smoking crater. Enemy position obliterated.' He paused for a second to let it sink in. 'Repeat: enemy obliterated. And sir, I gotta bug out, 'cause I'm all out of fuel.'

Fair enough. *Enemy obliterated*. What more could I ask of him?

FOUR
RIPPED

The F-16 got ripped by a pair of F-18s, which I'd have on station for two hours. It was 1445 by now, and the 2 MERCIAN lads were on the move again, pushing further into enemy terrain. But we now had a barrage of mortars smashing into the Green Zone.

From the Vector's open turret I could see the smoke plumes of those explosions. The mortars were impacting four hundred metres in front of us, and two hundred behind our forward line of troops. The barrage was creeping closer to our lads, and it wouldn't take long for the dicker to walk the enemy mortars on to target.

I split the F-18s. I got *Devo Two Two* over a two-mile-square grid where we reckoned the mortar team were firing from. I briefed the pilot to search with his FLIR (Forward Looking Infra Red) scanner for a hot mortar tube. If he found it he was to smash it.

I got *Devo Two One* over the Green Zone to the front of our line of troops. All three platoons were in fierce contact now, sandwiched between the enemy to their front and a mortar barrage at their backs.

The focus of enemy fire seemed to be coming from a patch of dense bush two hundred and fifty metres to the north-west of our lads. I gave *Devo Two One* the coordinates of a hundred-metre-square box to search. Within minutes the pilot came back to me.

'Visual six pax two-two-five metres north-west of your lead platoon. Visual four pax with weapons. Visual with muzzle flashes all along the woodline.'

'Nearest friendlies 225 metres south-east of enemy,' I told the pilot. 'Describe enemy position.'

I needed a better idea of the target, so I could work out how best to hit it. Our lead platoon were close to the splinter distance of some of the weapons that the F-18 was carrying.

'Six pax have taken cover in a narrow ditch in the woodline,' the pilot replied. 'Visual with muzzle flashes from out of that ditch position.'

'Right, I want you to drop a GBU-12 airburst right on top of 'em,' I told the pilot. 'Attack line coming in from the south-west to north-east. Confirm.'

The pilot repeated the details back to me. Coming in on that run he'd be flying over the heads of our lads as he launched his strike. But the trajectory of his attack should throw the blast away from our forces, or at least that was the theory.

A GBU-12 is an eight-hundred-pound smart bomb that can be set to 'airburst' mode, meaning it detonates one hundred metres above the target. It sends its explosive force downwards in a funnel of shrapnel that follows the bomb's momentum. It was the only way to hit those enemy fighters in that ditch, and keep the blast away from our lads.

I listened in as *Devo Two One* warned his wing of his attacking run, to deconflict the air, and then he gave me the sixty-seconds call. But as Sticky went to pass the warning to the platoon, there was the scream of an incoming mortar.

Sticky and I dived into the open turrets, but we were too slow. An instant later there was a crunching impact, the round smashing into the dirt not sixty metres from our wagon. The wave of the explosion tore across us, and I felt the stinging pain of blast-driven dust and rock and shit smacking into me.

But I was halfway through doing a live run with an F-18, and I was the JTAC who was calling the bomb: I didn't have the time to worry about getting hit.

'Time to fucking man it out!' I yelled at Sticky.

—

We let out a demented cackle, and thrust ourselves back out of the armoured turrets of the Vector. I swivelled and searched the skies to the south-west for a glimpse of a speeding F-18 Hornet. Almost immediately I spotted the gleaming dart of the aircraft on the far horizon. The pilot was right where I wanted him.

Let's get the bomb in.

'Call for clearance,' came the pilot's voice.

'No change friendlies. Clear hot!'

'In hot.' A beat. 'Stores.'

The GBU-12 is three metres long, and it 'flies' on a set of tail wings. It can be released from several kilometres away, gliding into target with a nine-metre margin of error. At a cost of some $20,000 it was far from being the most expensive munition in the F-18's arsenal, but it was a peachy one.

Released at height and distance it could take a good thirty seconds to reach target – plenty of time for the 2 MERCIAN lads to get their heads down. This time, there was no conventional ground explosion. As the GBU-12 detonated, the sky above the Green Zone erupted in a massive ball of raging fire.

The blast tore downwards from the epicentre of the explosion. Fingers of hot shrapnel rained on to the enemy position, throwing up a plume of dirt and debris where they smashed into the earth. That enemy ditch position had to have been smashed, but still I needed a BDA.

'BDA: there's nothing left alive down there,' came the pilot's voice. 'Correction: one male pax crawling away from the blast site.' A beat. 'Correction: he's stopped moving. Unsure of how many killed, but there are tiny heat spots everywhere.'

'Tiny heat spots' equalled body parts. The six enemy fighters in that ditch had been shredded, along with anything else caught in the airburst's downblast.

Devo Two One had to break off and head for the refuelling tanker. I pulled *Devo Two Two* in over the lead platoon, and gave him the grid of the most forward troops. The pilot confirmed he was visual with our lads, and happy with their route of advance. He told me that he was scanning the terrain up ahead for any sign of the enemy.

As the F-16 went about its work, Sticky held up an Army-issue Yorkie bar. He traced the distinctive red and yellow wording printed on the metallic blue wrapping.

'Yorkie!' he drawled, putting on a deep and manly voice as he did so. 'It's not for civvies!'

I couldn't remember the last time I'd had any scoff. I grabbed the proffered bar, tore off one corner with my teeth, and sucked the molten chocolate down in one greedy blast. Everything melted in the intense Afghan heat: food, shoes, your brains even. This was the only way to eat a Yorkie, plus it gave the body an instant burst of energy.

There was a squelch of static in my TACSAT. '*Widow Seven Nine, Devo Two Two.*' There was an urgency in the pilot's voice. 'Tell your lead platoon to go firm! Repeat: your lead platoon to go firm.'

I flicked my eyes across to Sticky, knowing that he was monitoring the air net. He gave me a nod, and put the call through to the OC. Not a word had been spoken between us. That instinctive communication was all part of the joy of conducting the ground-to-air war.

'Roger. Lead platoon going firm,' I confirmed to the pilot.

'I'm visual four males going into the treeline three hundred metres ahead of your lead troops. They're taking up positions on the track along which your men are advancing. I now have six pax spaced thirty metres apart, visual two with AK-47s.'

'Roger that. Wait out.'

I asked Chris to confirm with the OC that we could attack. They weren't firing at our lads, but they had been PID'd with weapons, and they were in ambush positions on our line of advance. The OC came back saying he was happy for the strike to go ahead.

For a second I considered what weapon to use. The F-18 carries an M61 Vulcan cannon, so maybe a strafe would do it. But the enemy were well spread out in a 150-metre stretch of dense woodland. The F-18's six-barrel 20mm cannon wasn't quite the A-10 Warthog's seven-barrel 30mm Gatling gun. Instead, I opted to go for bombs.

'*Devo Two Two*, *Widow Seven Nine*. I want immediate attack on target using two GBU-38s, coming in on a north–south attack run.'

'Affirmative. Two GBU-38s dropped simultaneously on target.'

I cleared him in to attack, and he gave me the 'in hot' call, the last before 'stores' – bombs away. Before he was able to release, Chris spotted the plume of a mortar firing in the far distance. *At last: we were visual with that bastard enemy mortar team.*

Chris gave an 'all stations' warning of the F-18 bombing run, so all ground call signs could get their heads down. He also warned the OC that he was visual with the mortar firing point. He reached for his map, and began trying to work out the grid from where the mortar was firing.

I got the 'stores' call from the F-18 pilot at the same moment that the OC came up on the net, telling us to smash that mortar tube – for under the rules of engagement we had every right to do so. The F-18's bombs were in the air, and there was nowt I could do but wait for the impact. So I dialled up *Devo Two One*, the F-18's wing.

'*Devo Two One*, *Widow Seven Nine*, sitrep: enemy mortar located three kilometres to the east of our position. Just fired, so tube will be hot.'

'Roger. Fully refuelled and two minutes out of your ROZ. Just as soon as I'm in the overhead I'll start my search …'

The pilot's last words were lost as a massive double blast roared across the valley: BOOOM-BOOOM! Two GBU-38s had ploughed into the earth one after the other, smashing apart either end of the woodline.

Each threw up a boiling plume of debris, from out of which an angry cloud of dark smoke billowed skywards. As the explosions reached their zenith they merged into one giant wall of searing grey-black stretching all along the woodstrip.

'*Devo Two Two*, BDA. Wait out.'

I wanted a battle damage report from the pilot, but first I had to control the jet searching for the mortar.

'Jackpot!' Chris exclaimed, as he passed me back a scribbled note of the mortar's grid.

'*Devo Two One*, *Widow Seven Nine*, I have enemy mortar grid: 46278190. Repeat: 46278190. Readback.'

The pilot repeated the grid.

'Affirm,' I confirmed. 'I want you to find that base plate and smash it.'

'Roger. Two minutes out from ...'

'*Break. Break*,' his wing aircraft cut in, using the codeword to clear the frequency of all traffic. 'BDA: four pax dead. Low fuel. Tanker.'

The brevity of the pilot's message said it all. He was sipping on air and breaking off for an urgent refuelling.

Chris briefed the OC that he would lose air cover for several minutes, as we had one F-18 refuelling and the other searching for the mortar. There was another distant bang and a plume of smoke. It was dead on the grid that Chris had given for the mortar.

To formulate a grid from a visual reference point is about the hardest thing in our game. Chris would've checked out the terrain as he could see it nearest the mortar, and chosen a couple of distinctive features – maybe an odd-shaped compound or distinctive hillock. He'd then have matched those with what he could see on the map, and worked out the grid from there. Chris was a bloody genius at it. The best I'd ever seen. I'd never known him get a single digit wrong. And he was bang on this time.

'*Devo Two Two*, *Widow Seven Nine*,' I radioed the F-16. He was still a minute out and I wanted to refine the plan of attack. 'Bank up to 30,000 feet, and don't come below. I want you to search around that mortar grid and tell me what you see.'

'Roger. Climbing to 30,000. Zooming in my optics to grid as given now.'

With one F-18 having left the airspace, I didn't want the mortar crew to know I had another jet coming in. At 30,000 feet the F-18 would be totally silent and invisible. That mortar was the single greatest threat we faced right now: it was targeting us and, more importantly, the lads in the Green Zone.

Ninety seconds later I got the call that I was waiting for. 'Sitrep: at grid given I see three males standing around a glowing metal tube. And guess what – I've just seen 'em reloading it.'

Chris radioed an all-stations warning that a mortar round was about to go up, so the lads could get into some good cover. There was a distant boom, and the pilot radioed me that he'd just watched the muzzle flash of its firing.

'Confirm no civvies in the area,' I asked the pilot.

'Affirm. No other pax present.'

The enemy were renowned for sighting their mortar tubes with women and children gathered around them, as cover. I had to double-check and brief the OC. Ultimately it was his call, but one that he'd delegate to me.

'*Devo Two Two*, hit it as fast as you can any line of attack,' I told the pilot. 'Your choice of ordnance.'

I gave him final clearance and he gave me 'stores'. We were all eyes on the far horizon. There was a sudden flash, followed by a boom, and a couple of seconds later a mushroom plume of smoke rose into the distant sky. He'd hit it with a 500-pounder, I reckoned.

'*Devo Two Two*; BDA.'

'It's a Delta Hotel,' came back the pilot's reply. Direct Hit. 'There's bits of warm pipe everywhere. And nothing left of the three pax around the tube.'

Fucking result.

It was 1630 by now, and we'd been in the game for eleven hours solid. Unbelievably, we'd yet to take any casualties. The platoons were just short of the three targets – Objectives Silver, Gold and Platinum – and the limit of their advance. They'd been bar-mining their way into compounds, blowing holes in the walls and clearing them as they went.

The bar-mines were hammered on to the wall with spikes, and the flick of a switch set off a fifteen-second fuse. There'd be the cry of 'MINE!' Then the crump of an explosion. As soon as the hole was blown, the lads would follow through with grenades. We didn't know which doors and entrances might be booby-trapped, so the only 'safe' way in was by blowing the walls.

As each new patch of territory fell to us the radio chatter was going wild, with enemy commanders urging their men to stand and fight. It was far from over yet.

The two jets were ripped by a singleton F-18, call sign *City Desk Four One*. I was getting shedloads of F-18s launched off an American carrier steaming in the Gulf. It was all good by me. The American pilots were doing sterling work of smashing what I told them to smash, whenever I told them to smash it. It was a top job.

As I talked the new pilot around the battlefield, the lads of the 2 MERCIAN sniper team came over to have a natter. They'd been up on the high ground all day long, but hunkered down in their hides. They'd seen little or no action, for most of the contacts were happening at the far end of their effective, eight-hundred-metre, range. I was feeling a little sorry for them.

The two lads looked to be no more than eighteen- or nineteen-years old, and they carried these long, L96 sniper rifles. We shared an Army ration milkshake, my other favourite scoff when in continuous action. I'd kept a couple of water bottles on the burning roof of the Vector, the contents of which were the perfect temperature for dissolving the powdered shake.

From the turrets we had a good vantage point over the battle-field. As we supped our shakes and gazed out over the Green Zone, I ribbed the sniper lads about how they should have trained as JTACs. We spotted movement some eight hundred metres away, in territory where we'd just been smashing the enemy.

I was about to alert the F-18, when one of the young lads took a butcher's through his scope. The L96 is fitted with a Schmidt & Bender 12x magnification sight. He had two enemy figures in the crosshairs of his scope. Both were armed, and they were advancing towards our troops.

I watched in fascination as this teenage lad flipped out the bipod of his weapon, and settled himself down to fire. He squeezed off the first shot, adjusted his aim, and squeezed off a second. We were just about to congratulate him – two shots: two kills – when all hell broke loose below us.

Pushing up towards the main target – Objective Platinum – the lead platoon had stumbled into another hornets' nest. They had machine-gun rounds and RPGs slamming into them from a treeline just to their front.

Major Butt was on the air immediately, requesting a danger-close air mission to smash that enemy position. Their fighters were posi-tioned around 100 metres ahead of our lads, and they were pushing men forward to surround and outflank us.

I talked *City Desk Four One* on to the enemy in the treeline, and told him to look a hundred metres to his west for the lead platoon. When he was visual with our lads I told him I needed a danger-close strike to smash the enemy. I asked him what ordnance he'd recom-mend at a hundred metres' distance from friendly troops.

'A thousand-pound JDAM,' came back the pilot's calm reply.

It wasn't quite the answer I'd been expecting. A thousand-pounder was twice the weight and destructive power of anything I'd dropped so far, yet this was the most danger-close air mission.

I swallowed hard. It was the JTAC who bought the bomb, and I knew that I'd never be able to live with myself if I smashed my own lads.

'A thousand-pounder?' I queried. 'Not owt a bit smaller?'

'Sir, that's a pinpoint-accurate munition,' came the pilot's reply. 'As long as your boys have their heads down, they'll be OK.'

I flicked a glance at Sticky. He gave me a thumbs-up. There was something about the calm tone of the F-18 pilot that gave me real confidence in his abilities.

'Roger, a thousand-pound JDAM,' I confirmed. 'Attack on north–south run, to keep the blast away from friendlies.'

The pilot told me he was tipping in, and called for clearance. I had him visual to the north of us, and I could tell he knew what he was doing. I could hear Chris screaming into the radio for all stations to get low. I gave the pilot the green light.

'You're clear hot. Ground commander's initials are SB.' 'SB' for Major Simon Butt.

'In hot,' the pilot confirmed. 'Stores.'

A moment later I heard the faint whistle of the incoming JDAM. Within seconds it grew into an ear-piercing scream. The noise was like an express train speeding down a tunnel with us at the very end of it. It drilled into my head. There was the flash of a wheelie bin-sized object streaking through the air in front of us, and then the thing hit.

The blinding flash of the explosion burned away the late-afternoon shadows, leaving a fuzzy white blob on my retina. The sound and blast wave hit, rolling and thundering across the valley in a deafening tidal wave of noise, rocking the wagon backwards and forwards on its suspension.

Sticky and I gazed in open-mouthed amazement at the impact point. The boiling cloud of dust and debris tore ever upwards and outwards, dwarfing the bush and the compounds that lay below it. Chunks of masonry and trees flew high into the air, each trailing its own dark and angry finger of smoke. From the centre of the explosion an ink-black pillar of burning barrelled into the air, pooling into a mushroom cloud high above the strike point. Unsurprisingly, the battlefield had fallen echoingly silent.

I turned to Sticky to get a sitrep from the lead platoon. As I did so rounds started coughing out of the treeline, one hundred metres to the north of the JDAM's impact point. I could barely believe that anyone was left alive in there – but they were, and they were still fighting.

The OC was up on the net immediately: 'Bommer, we need that target sorting! Get the air in again now!'

'*City Desk Four One, Widow Seven Nine,*' I yelled into the TACSAT. 'I want immediate re-attack on the same target, but one hundred metres north. Put a five-hundred-pound airburst over that position.'

'Roger that. Banking round.' A beat. 'Tipping in now.'

The pilot brought the F-18 around in a spanking turn, streaks of cloud-like vapour trail clinging to the twin, v-shaped fins of its tail. Forty seconds later I cleared him in hot. He put the airburst exactly where I'd asked, and the northern end of the treeline was torn to shreds in the blast that rained down from the air.

The pilot was low on fuel, and he got ripped by *Devo Two One* and *Devo Two Two*, the pair of F-18s that I'd had earlier. It was 1800 by now, and three platoons were on their objectives. They were a thousand metres away across the Green Zone, and in the gathering dusk I couldn't see much with the naked eye.

I got the pair of F-18s flying reccees over the objectives, but no further enemy fighters were seen. By 1830 Objectives Silver, Gold and Platinum had been taken. Whether it was the air power or whatever had done it, the enemy's will to fight seemed to have been broken.

They'd bugged out leaving behind three huge, mud-walled compounds stuffed full of ammo, weapons and big bales of opium. There were also stacks of maps, notebooks and other useful Intel. The platoons went firm and set about destroying all the weaponry they could find.

—

As darkness fell across the valley, the men of 2 MERCIAN began their withdrawal. The mission brief called for all friendly forces to be out of the Green Zone by nightfall, and laagered up in the comparative safety of the open desert.

But as the men fell back through the silent territory they'd just been fighting across, the rear platoon got hit. All of a sudden I could see the fiery trails of RPGs and tracer rounds sparking red through the thickening Afghan night.

I got the F-18s overhead the contact point. Almost immediately *Devo Two One* picked up two glowing, pipe-like objects – the heat signatures of RPG launchers that had just fired. I asked the pilot to hit them with a GBU-12 in non-airburst mode.

As the eight-hundred-pound smart bomb smashed into the position, there was a burst of white-hot fire that lit up the entire night sky, fading to a darker orange at the edges. Walls and trees and rooftops were silhouetted in the heat of the explosion, which fired the valley a ghostly volcanic red.

The OC's voice came up on the net. 'Cheers, Bommer. Thanks for that.'

I asked for an immediate BDA, and the pilot reported that the heat spots had gone. Finally, all had fallen utterly silent across the night-dark battlefield.

The platoons withdrew past the ridge line and pushed into the desert, and we prepared to leave our position on the high ground. As Throp gunned the Vector's motor and turned the wagon away from the battlefield, I presumed I'd seen the last of Adin Zai. But in fact, this very stretch of terrain was to become our permanent battleground. We would be back. Today was just one day, and we would spend the next hundred days fighting here.

And when we returned, the enemy would be waiting for us with a bloody vengeance.

FIVE
TAKE US TO THE BODIES

Our convoy of vehicles was parked in the open desert, with a skeleton crew as security. We rejoined them, and laagered up in all-round defence. I checked with the platoon commanders, and not a single man had been injured. It was an incredible result, after thirteen hours of intense combat at close quarters.

The lads gathered in the safe harbour created by the circle of armoured vehicles, and started throwing around an American football. We got a brew on using an empty 7.62mm ammo tin and some hexy solid-fuel blocks. You could get six good Jack flasks (Army-issue metal cups with screw-on lid) out of one ammo tin, so there was more than enough for the four of us in our FST.

Like every proper north-east of England lad, I can drink tea until it comes out of my ears. I'd been dying for a good brew all day long. But I'd barely taken my first sip when a call came through on the TACSAT. I had an F-18 inbound, call sign *Voodoo Five Two*. No rest for the wicked, I told myself, as I briefed the pilot on what I wanted doing.

I got him flying recces around the perimeter of our laager, and he reported the terrain as deserted. Then I tasked him to fly some recces over the battlefield. All he could see were a couple of tractors and trailers pottering about in the Green Zone. I asked him to take a close look. A few moments later the pilot was back on the air.

—

'*Widow Seven Nine, Voodoo Five Two*. Those trailers are stacked high with bodies. I guess that's the enemy haulin' out their dead and injured.'

Keeping one ear on the pilot's commentary, I reached for the ratpack that Sticky was holding out to me. Over the past few weeks Sticky had taken it upon himself to be the FST's honorary chef. He'd chucked four of the silver foil-clad heat-in-the-bag meals into the ammo tin, and boiled up some scoff.

I ripped off the top and stuck my nose into the steaming bag. Ah – lovely! Meatballs and pasta. I grabbed my spoon, which I kept jammed in the top of my radio pack, and dug in. When I'd finished eating I cleaned the spoon by giving my brew a good stir, then jammed it back in my pack.

I was fed and watered and dying for some kip, but I still had that F-18 on station. Keeping a listen on the pilot's commentary, I pulled out my JTAC log, and did the next vital task. At the end of every battle the JTAC is supposed to submit a mission report ('missrep') on every live drop – a JTAC-controlled attack using an air asset.

One of the main reasons for doing those missreps is in case of friendly fire or civilian casualties. As every JTAC knows only too well, if we dropped a bomb or did a strafe and killed some of our own men, we would be held legally responsible. Likewise if we killed some Afghan civilians who had somehow wandered on to the battlefield.

Since leaving FOB Price at the start of the operation I'd done 115 air controls, so there were a good few missreps to write up. I scribbled away, my head torch casting a faint halo over my notebook – black pen for non-use of munitions; red for live-fire missions.

I ran through the missrep headings that I'd learned back in JTAC school: bearing; distance; target location (lat & long); target elevation; target description; attack heading; friendly forces; hazards; weather (if significant) … I tried to stifle a yawn.

Major Butt came over for a chat, which was a good excuse to break off what I was doing. He was a gruff, tough kind of commander, and not the sort of guy who gave praise lightly. The word was that the OC had been a professional rugby player in his youth, and he certainly had the size and the physique for it. I reckoned the guy could give Throp a good run for his money.

'Bloody cracking op,' remarked Butsy. He seemed in an unusually talkative mood. 'Couldn't have gone better. Everything went as planned. How about from your end?'

'Aye. Top op, sir,' I confirmed.

'It all went without a hitch, sir,' Chris concurred. 'From the FST's perspective, not a single problem with the guns or the air.'

Chris was actually the second most senior rank in the company after the OC. The mission plan allowed for him to take over command, if the major got injured or otherwise taken out of action.

'Still, let's not underestimate these guys,' Butsy remarked. 'Look how swiftly they reacted to us being on the ground. As we were massing for the op those black-clad figures were forcing women and children out of the village. They pushed their fighters forward, and got the civvies out. And you saw the sophistication of their dicking procedures? They had guys on the high ground flashing with mirrors and torches all around us.'

I took a slurp of tea. 'Aye.'

'There was one moment we saw them looking through their binos,' the OC continued. 'I had this instinctive sense of let's not move forwards, and ordered the lads to stop. In that instant three RPGs flashed in front of us. If we hadn't stopped the four of us would've been whacked. There was this voice in my head that told me to stop, and the RPGs flashed in front of our bloody noses. I reckon we make our own luck, but that was the first time I realised they were targeting the HQ element specifically.'

'They were?' I let out a half chuckle. 'I guess that explains why it was you lot kept getting smashed. How many times did I put up the call – "HQ element surrounded and getting smashed …"'

The OC grinned. 'There we were lying in the dirt, and eventually the penny dropped: *they're trying to take out the HQ.* That's how smart they were …'

For a while I lay on my back half-listening to the OC, and gazing up into the wide expanse of the burning, starlit sky. Then Sticky came to have words.

'Bommer, someone's trying to raise you on the air.'

'Who, mate?' I asked. 'What's his call sign?'

'Fuck knows,' he shrugged. 'He won't tell me. Says he wants the JTAC.'

I grabbed the TACSAT. 'This is *Widow Seven Nine* for any call sign in my ROZ.'

'*Widow Seven Nine*, good evening, sir,' came back the unmistakeably American voice. 'This is *Tin Can Alpha*.'

I nearly choked on my brew. '*Tin Can Alpha*!' I spluttered. 'Are you winding me up?'

'No, sir. That's our call sign, sir: *Tin Can Alpha*.'

I'd never once heard of the call sign *Tin Can Alpha*. It had never been mentioned, not even in the briefings I'd received at Kandahar airfield at the start of the tour.

'Well, what kind of platform are you, *Tin Can Alpha*?'

'You don't need to know that, sir,' the voice replied. 'We're an American airframe in overwatch your position. Sir, I'm tasked to ensure that you don't get ambushed down there.'

I felt a horrible sinking feeling. 'Erm … right-oh, *Tin Can Alpha*, how long do I have you for?'

'We have five hours' playtime, sir.'

Oh shit. I glanced at my watch. It was 2300 hours. That meant I'd have this wanker until 0400, whilst everyone else was getting their brackets down. I was gutted. Three hours later *Tin Can Alpha* sounded bored shitless with flying orbits over a deserted patch of desert. But I'd bet my bottom dollar he wasn't as bored, or as knackered, as I was. I switched from tea to coffee in a desperate effort to stay awake. As a way to kill time the pilot started asking me all about the battle for Adin Zai. In return, I started asking him about the capabilities of his top-secret aircraft, but he wasn't telling.

'I'm sorry, sir, I can't tell you that,' he kept repeating. 'It's classified, sir.'

It was like he was on permanent replay. I loved the American attack-jet pilots, and I loved the American warplanes – it was just *Tin Can Alpha* I could have done without. Somehow, we reached 0400 with me still awake, and *Tin Can Alpha* finally signed off the air.

'You stay safe and happy down there, sir,' were the pilot's closing words.

'Aye,' I replied. 'I'll try to, mate.'

I felt like adding – *as long as I don't get sent any more platforms with idiotic fucking call signs.* Instead, I got my head down on the sand and was instantly asleep.

An hour and a half later someone was shaking me awake. I'd been kipping on the deck, curled up against the wheels of the Vector.

'Bommer. *Bommer.* You've got air.'

It was Sticky. I was dog-tired, and I didn't say a word. I grabbed the TACSAT, plus the fresh brew he handed me.

It was 0530. First light. A sickly-yellow sun was clawing its way above the low mountains to the east. I had a Harrier GR-7 above me, flown by one of the pilots I'd been working with at the start of the previous day's action.

'How're you doing?' he asked, once I'd finished the area of operations update. 'I hear you've been busy down there.'

I told him I had, and that it was all-good. He stayed with me until 0730, checking out positions around Adin Zai, but nothing much was moving. Then he signed out of my ROZ low on fuel.

'Watch yourselves,' the pilot told me, 'and good luck with the rest of the op.'

I dozed a little, but it was useless. Everyone was awake and making a racket, for the entire company had done stand-to at first light. I gave up and decided to give my teeth a good scrub. Then I went and helped Sticky get the breakfast on.

At 0800 the OC set off for the Green Zone, to rendezvous with the elders of Adin Zai. His psyops (psychological operations) team had arranged for a *shura* – an Afghan powwow – with the village chiefs, the aim of which was to explain exactly what yesterday's shit-fight had been all about.

Basically, we'd been in there and smashed the enemy, and it was the gentlemanly thing to make clear why. We'd targeted a known Taliban stronghold, and only fired upon when fired at – and that's what the OC would explain to the elders. If they understood why we'd been fighting our way through their village, it should help keep them onside.

Apart from his HQ element of four men, the OC had two platoons with him for security. As he headed to the *shura*, Major Butt planned to leave men guarding the route out, in case of any foul play. That way, they should be able to fight their way through any ambushes and extract if they needed to.

Sticky, Throp, Chris and I loaded up the wagon, and set off for our favourite position. Butsy wanted air on hand, as a deterrent to any nonsense at the *shura*. I had two F-18 Hornets check into my ROZ, call signs *Wicked Three Three* and *Wicked Three Four*. Apart from the tractors and trailers which were still at work hauling out the dead, nothing much was seen by the pilots.

The OC made the rendezvous with the villagers without incident. A crowd of old men with turbans, and younger men with beaded skull caps gathered, whilst the OC stood out front and addressed them via a 'terp', or interpreter.

The elders reported that thirty enemy fighters had been killed. As a result of the battle, the entire Taliban presence had been forced out of Adin Zai. The OC radioed us the good news, and warned us that he was starting his push back towards the laager. Once he'd rejoined us, we would depart for FOB Price the way we'd come in.

Major Butt had barely left the *shura* when it kicked off big time. From the Vector's open turret I could hear the repeated crunch of RPGs and the long bursts of small arms fire. The OC radioed us that his HQ element had been hit in a double-sided ambush. He was hopelessly outnumbered.

From the west, the men of the platoons were trying to fight their way through to relieve him. The enemy forces were 140 metres north and east of the OC's position, and closing fast. This was danger-close, but it was nowhere near as tight as some of the air missions I'd done during the previous day's battle.

I radioed the F-18s.

'*Wicked Three Three*, *Widow Seven Nine*. Sitrep: I've got my HQ element in the Green Zone being hit hard. Friendly coordinates are: 62903781. Readback.'

The pilot confirmed the details. I explained that I had enemy forces one-four-zero metres to the east and north of the OC's position, and asked the pilot to smash them.

'I want immediate attack using GBU-38s, on a north-east to south-west run. You're danger-close to friendlies. Ground commander's initials are SB.'

'Roger that,' the pilot confirmed. 'Programming one GBU-38 to drop on each of the enemy positions. Banking around. Call for clearance.'

I watched the pilot tear about in a screaming turn to bring the F-18 on to my line of attack. I was urging him to get a bloody move on. What a shit state we'd be in if the OC and his lads got killed or captured on the day after the battle – especially as all he'd been doing was having a chat with the locals, to win some hearts and minds.

Suddenly, the F-18 pilot was back on the air. But all I could hear was the horrible rhythmic wailing of warning alarms blaring away in his cockpit.

'I got big problems up here,' the pilot yelled above the racket. 'Breaking off my attack. Returning to base. *Wicked Three Four* on task and awaiting your call.'

That was it. He aborted the attack run and signed off the air. It was fair enough. From the sound of those alarms it was like his jet was about to fall out of the sky. I radioed his wing, and repeated the attack instructions. The pilot said he needed a minute to get into a position. I was cursing to myself as I counted down the seconds to clear him in.

There was a squelch of static and I knew immediately that something was wrong. The pilot was in the midst of a screaming turn, and he couldn't be calling with any good news.

'*Widow Seven Nine, Wicked Three Four* – I've got a total weapons computer malfunction. I am unable to attack. Repeat: unable to use any weapons.'

That was both F-18s out of action, and still the OC was deep in the shit. This was getting desperate. I asked *Wicked Three Four* to fly a show of force over the enemy positions at fifty metres altitude, firing flares.

'I want repeated shows of force,' I yelled at the pilot. 'And as low as you can make them.'

'Affirmative,' the pilot replied. 'Starting shows of force now.' 'Wait out.'

I flipped frequency on my TACSAT, to bring me on to that of *Stoneage*. *Stoneage*, based at Kandahar airfield, is the top dog in terms of all air missions in southern Afghanistan. As JTACs we were only ever supposed to call *Stoneage* in an emergency. With two F-18s out of action and the OC deep in the shit, this was it as far as I was concerned.

'*Stoneage, Widow Seven Nine*, d'you copy?'

As I waited for the reply, I felt like I used to when waiting to see the headmaster after causing trouble at school. But fuck it, this was dire and I had to do something.

'*Widow Seven Nine*, this is *Stoneage*,' came the gravel-voiced response.

'Sitrep: I've got my HQ element surrounded and at risk of being overrun in a danger-close contact. I have two *Wicked* call signs out of action with systems failures. I need immediate G-CAS. Repeat: immediate G-CAS.'

'*Widow Seven Niner*, I have pilots running for the fast jets to launch G-CAS now,' came the reply. 'I'll radio in time-to-target once they're in the air.'

G-CAS stands for Ground-launched Close Air Support. It was the quickest emergency air cover available when no other fast jets were free and in the air. The main advantage aircraft like the F-18s, F-16s and F-15s have over the Apache – apart from the heavy ordnance they can carry – is their time from launch to target.

With a top speed in excess of Mach 2 at altitude, and a rate of climb of some 17,000 metres per minute, an F-18 could reach us in the fraction of the time it would take an Apache gunship. It made them worth every dollar of the $30 million it cost to build one.

By now *Wicked Three Four* had flown three shows of force. From the open turret of the Vector, I'd seen the last go in at what looked like twenty-metre altitude. But it was having bugger-all effect on the enemy. Their commanders had obviously learned the lesson from yesterday's battle: shows of force meant little, and they were to keep attacking.

The OC reported eight enemy kills, but still he was being hit by a murderous barrage of small arms and RPGs.

'We're in the dirt, well isolated and it's not looking good!' he was yelling on the radio, to Chris. 'Tell Bommer we need something now!'

Where the fuck was that G-CAS? As if in answer, there was a squelch of static as the big man came up on the air.

'*Widow Seven Nine*, *Stoneage*. You've got two *Uproar* call signs scrambled, inbound to your position.'

Just as soon as the pair of F-18s had checked into my ROZ, I passed *Uproar Two Three* the coordinates of the enemy positions and cleared him in to attack. The F-18 pilot zoomed in his optics to the coordinates, and immediately he was back on the air to me.

'Visual enemy position. Visual ten to twelve pax in the wood-strip, with muzzle flashes.'

'I need immediate attack with an airburst munition on the centre of mass of enemy. Attack line north-east to south-west run.'

'One minute out,' the pilot confirmed. 'Tipping in. Call for clearance.'

I cleared the F-18 pilot to attack, and he released a GBU-38 airburst. The explosion ripped apart the air above the enemy posi-tion, and tore into the woodland below, hurling branches and chunks of earth high into the air.

'Get in!' I yelled. 'I need BDA,' I radioed the pilot, hoping to god I hadn't smashed any of our lads.

'BDA: seven pax KIA in the treeline,' the pilot replied. 'I can see survivors fleeing their positions, and running away from your forces.'

There were seven killed in action (KIA) that the pilot could see, and probably a whole lot more that he couldn't. The enemy were on the run and had been broken. There was no need for a follow-up attack. Butsy and the men of B Company had survived again, and it was time for them to get the hell out of there.

We regrouped at the laager and the convoy began forming up for departure. But this was when it all went totally warped. Before we could set off, the Mortar Company Commander came to have an urgent word with the OC. He had with him one of the terps, and as we gathered around they related a simply incredible story.

At the start of the previous day's action, a young Afghan male of fighting age had blundered in to the Mortar Company's position. The young man was shouting like a madman, but being a mortar company they had no terp with them. They threw the guy in the back of the 'greeny wagon' – the Mortar Company truck – until a terp could be found to talk with him.

Apparently, that had just happened, and this was the young man's story. He claimed to be one of ten Afghan policemen who had been stationed at Zumbelay, a town to the east of us across the Helmand River. Six days ago the Taliban had kidnapped him, and nine other policemen, from the local cop shop. The Taliban had taken the ten men to Adin Zai. There they were stripped of their boots and uniforms and held captive in the village mosque. Three days later us lot had pitched up on the desert horizon, massing to attack Adin Zai.

The mosque was four hundred metres from our line of departure, and it was from there that we'd received the fiercest resistance as our initial assault went in. The enemy had concluded we were trying to rescue the kidnapped policemen. They'd bundled the ten men out of the mosque and driven them into the desert.

The Taliban had taken them to a sunken wadi, and ordered them to make a run for it. As they'd sprinted for their lives the Taliban came after them in pickups, hunting them down. I guess that was their idea of a bit of sport – gunning down unarmed policemen. In the ensuing mayhem the young man had escaped.

He'd run across the desert for two hours solid, before blundering into our position. The poor guy's feet were torn to shreds. What gave his story added credibility was this: we'd heard about the kidnap already. Four days prior to the start of the present mission it had been the main item of interest in the Intel brief at FOB Price.

The question was – what did we now do about it? The OC put a call through to the Commanding Officer of 2 MERCIAN, Colonel Richard Westley. Butsy and the lads were totally shattered, and looking forward to returning to base. But it wasn't to be.

The response that came back from the CO was this: if at all possible we were to go and recover the corpses of the murdered coppers.

And so we put the young Afghan lad in one of the Vikings, and told him to take us to the bodies.

SIX
THE SOMME

We drove for just over a mile across the baking desert until we found the first corpse. The young Afghan lad had led us to the lip of a wadi. We pulled up on a stretch of high ground that seemed to dance and shimmer in the heat. This was where the Taliban had offloaded the ten policemen, he explained, before they started shooting.

We approached the edge of the dry valley on foot. The young policeman gestured over the edge, and started shouting and wailing and tearing at his hair. I peered over, and some ten metres below was the body. It was a young Afghan lad who looked to be no older than fourteen. He had a bullet hole in the centre of his forehead.

We followed the distraught Afghan policeman down the slope, at the bottom of which were three more crumpled forms. All of them had been shot in the head at close range. The telltale black burn marks of cordite were visible around the entry wounds, showing how close the gun muzzle had been when the round was fired.

One of the captives had been made to pray before the Taliban put a bullet in his brain. He was stone-cold dead and frozen in a gruesome caricature of prayer, still kneeling with his forehead on the dirt. Another had been shot through both his ankles, before taking a bullet in the forehead.

Three hundred and fifty metres down the wadi we found two further corpses. These guys had made a desperate bid to escape, and their bodies were riddled with bullet holes. The final three were scattered a way further down the wadi's dry bed, the last mile or so from where the Taliban had set them running. Tyre tracks in the sand showed where the Taliban gunmen must have come screeching after the terrified men, their weapons firing on automatic from the rear of their Toyota trucks. Their death run had been like a classic man-hunt, and I guess the Taliban had had their sick 'fun'.

A couple of the victims had been crawling on the ground in a desperate effort to escape, when the last of the bullets were pumped into them. The Taliban had chosen to do their torture and murder here, as the walls of the wadi would have shielded the noise of the gunfire from us.

The smell wasn't too bad yet, for it had only been a day and a half since the executions. But the palpable evil of what had happened here was sickening. Nine young policemen had been tortured and executed in cold blood. No one doubted the young survivor's story.

We returned to the high ground, and radioed a request for a team from the Special Investigations Branch (SIB) of the military police to attend the scene, plus the bomb-disposal boys. We needed the Ammo Technical Officer (ATO) lads to check that the bodies hadn't been booby-trapped. Prior to that, we couldn't begin to bag them up or move them.

We went back to the Vector, parked on the brow of the wadi. There was nothing for it but to get a brew on. Sticky threw some boil-in-the-bag meals in the ammo tin, and we had an early tea. It sounds harsh, but what else were we to do whilst awaiting the arrival of the specialists?

An hour later I got a call from an Apache gunship. The *Ugly* call sign was inbound to our position, escorting a Chinook transport helicopter. I got the gunship to fly a couple of recces around us, to check for enemy, as we were only five hundred metres from the Green Zone. Nothing was seen, and the Apache brought the big, twin-rotor helicopter in to land.

By the time it was dark the ATO and SIB lads still hadn't finished doing their stuff. On the original mission tasking we were supposed to have been back in FOB Price hours ago. Instead, we laagered up in the desert for a third night running, so we could finish dealing with the bodies come morning.

With our mission officially over I had no more air controls, which was a massive relief. Sticky, Throp, Chris and I were laid by the side of the wagon, with the dead bodies not more than twenty metres from us. But I couldn't let that unsettle me. I got my bracket down by the side of the Vector, and fell into a deep sleep.

I jerked upright to the sounds of an almighty explosion echoing through the night. At first I thought I was dreaming, but then I noticed the white-hot blasts lighting up the skyline. It was 2300 hours, and all hell had broken loose some thousand metres to the east of us. There was the juddering crackle of gunfire, the thump of heavier weapons, and the thunderous roar of repeated explosions rocking the desert air. I shook the sleep out of my head and tried to focus on what was happening. At first I presumed we'd been ambushed, but there didn't seem to be any of the fire hitting us.

For an instant I picked out the rhythmic thwoop-thwoop-thwooping of rotor blades, and the unmistakeable shape of an Apache gunship flashed in silhouette against the angry red of an explosion, as it banked around. What the hell was going on?

I grabbed my TACSAT: '*Widow TOC, Widow Seven Nine*. Sitrep: we're laagered up in the desert at Adin Zai. One click to the east of us there's a massive contact. There's an *Ugly* call sign in action and explosions and gunfire. What the hell's happening?'

'*Widow Seven Nine, Widow TOC*. No idea, I'm afraid. Wait out.'

Whilst Widow TOC asked around, I put out a message requesting any ground call signs to respond. No one answered. Next I tried this.

'This is *Widow Seven Nine* making an any-stations call. We're on the ground visual with a massive contact, and this is our grid: 90236784. I repeat: this is an any-stations call, *Widow Seven Nine.*'

There was a moment's echoing silence, then: '*Widow Seven Nine*, this is *Spooky Two Zero*. Sir, we're in the middle of an op, and you need to keep the traffic down.'

Spooky was the call sign of a specialised US airframe, one reserved for covert operations requiring immense firepower. That aircraft alone had the firepower to take out our entire convoy, and yet there was a mission going down that no one had bothered to warn us about.

'*Spooky Two Zero, Widow Seven Nine*, stop fucking about,' I rasped. 'We're laagered up less than a click away from your op and we need to know what the fuck's going on.'

'Well, I have this message for you, sir, just in: "Bommer, what the fuck're you doing out at this time of night?" Sir, that message is from Nick the Stick.'

Now I knew what was happening.

I'd befriended Nick the Stick back in FOB Price. He hung out down the American end of the base, and my main reason for going there was the grub. The US Army cookhouse would serve lobster, followed by ice-cream gateau, all washed down with chilled soft drinks. In the British mess tent you'd make do with bangers 'n' mash and a plastic cup of warm water.

Like most of the guys in the American base, Nick the Stick was a giant of a bloke. You could've fitted two of me into one of him, and still had room to spare. I guess his nickname – 'the Stick' – had to be a pisstake. All the US operators went by their first names only, and it didn't take a genius to work out what units they were from. But as all their operations were strictly classified, I wasn't about to go asking.

I had one card to play to blag my way into their mess tent: *Operation Silver*. Nick the Stick had heard all about that mission, and how I'd taken out a cadre of top enemy commanders in the one hit. I'd been on my first combat mission in Helmand, as the JTAC embedded with 42 Commando. We'd been ordered to take Sangin town on foot. My airstrike on the enemy commander's compound had earned me a certain kudos.

Nick the Stick was a JTAC newly arrived in theatre, and he'd wanted to know all about it. I was more than happy to fill him in, as long as the chefs kept piling my plate with lobster. I left the US mess tent with my combats stuffed full of scram – cans of Coke, Mars bars and the like. I didn't give a damn about the looks I got off the other American operators – *you sad, scruffy British bastard.*

Nick the Stick and I had bonded over the lobster, and after that we'd become good mates. And now I knew what the contact was to the east of us. It was a classified US operation going in, and Nick was calling in the airstrikes. I got on my TACSAT and made the call.

'*Nick the Stick, Widow Seven Nine.* Sitrep: we're a click to the west of you laagered up in the open desert. We're here retrieving the bodies of ten Afghan policemen murdered by the enemy. Now, what the fuck are you lot up to?'

'*Widow Seven Nine*, this is Nick. We're doin' a lift op, to the west of Adin Zai. And buddy, we got to keep the traffic down. I got your position, and we'll keep the fire away from your guys.'

I snorted. 'Cheers. And thanks for the early warning, mate.'

I briefed the OC and Chris. The US forces were doing an op under cover of darkness to snatch a high-value target – that's what Nick the Stick had meant by a 'lift op'.

Then I got another call from the aircraft above us. '*Widow Seven Nine, Spooky Two Zero.* It's going to get very noisy on the ground there for a while now ...'

'Thanks for telling us,' I cut in, sarcastically.

'Roger that, sir. And sir, if you're that close to the contact point you might want to have your men stand by to help our team extract, that's if we need you.'

The bloody cheek of it. First they'd launched a covert op right on top of us, and didn't bother to warn us. And now they were asking for our help, in case it all went to rat shit.

I told the OC what the pilot was suggesting. Butsy radioed the CO, who cleared us to assist. And so the entire company was placed on immediate notice to move, in case Nick the Stick and his buddies needed us to get them out of the shit.

The pilot then relayed a request from the task force doing the snatch operation. They were stood off in the desert, and they wanted us to launch a feint into the Green Zone. They were asking us to draw enemy fire, so they could sneak in and out again.

I briefed Butsy, and he passed it up to the CO. Having got clearance, Butsy now had to come up with an instant plan of action. From being exhausted and laagered up, it was flash-bang into launching a full-on combat assault to support the snatch operation.

Butsy put together a strike force, consisting of himself and his HQ element, plus 3 Platoon and our FST. He estimated we'd be up against no more than twenty enemy, after the pounding they'd taken over the past two days from us lot. We set off into the Green Zone, a line of troops navigating on night-vision and by the light of the moon.

We descended like a silent snake from the white of the moon-lit desert into the thick sea of darkened vegetation. I felt my adrenaline pumping. I loved missions like this one. It was what a soldier thrilled for – taking the fight to the enemy on foot and in a night-dark battlefield.

Overhead, we had a US warplane shadowing us, with its state-of-the-art scoping equipment keeping a watch for the enemy. We pushed a mile down the track that led into the heart of Adin Zai, when suddenly it all went pear-shaped. The bush all around us erupted in a wall of fire, as the enemy hit us with everything they'd got. Rounds went slamming into the undergrowth and the dirt track, and RPGs were churning overhead, trailing gouts of fire like giant spurts of lava.

As the bullets snapped angrily past, I felt a kick to my backpack. A round had pinged off the 'donkey dick' aerial of my TACSAT, going snarling past my head and burying itself in the bush. When in man-portable mode – ie, not in the wagon – the TACSAT went in my pack with a thick metal aerial, about the size of a donkey's dick, stuck out of the top. Hence the name. Luckily, it was made of thick, rubberised steel, and could take a few rounds.

The US warplane reported up to one hundred enemy fighters massing all around us, and closing fast. The OC decided the feint was most definitely over, and it was time to get the hell out of there. He was a man for doing battle at the time and place of his choosing. The trouble was, how on earth were we going to get out arses out of this one?

It was then that the pilot orbiting above us came up with a blinding suggestion. He proposed that we run hell for leather back the way we'd come, as he programmed his aircraft to hit the positions to either side of the track and to our rear with every weapon he carried. As we ran so he would shadow us, shepherding our progress with his awesome firepower.

The OC gave the order for all stations to retreat at full speed sticking strictly to the track, and we turned as one and legged it. As we did so the heavens opened up, the night sky below the invisible aircraft erupting in a seething fountain of white and blue flame that tunnelled earthwards, as first the cannons and then the bigger guns started to rain down fire.

An instant later the torrent of red-hot destruction tore into the vegetation to either side of us, which erupted in wild explosions. First came the small stuff, then the rhythmic thud-thud-thud of shells slamming into the earth to either side of us. Finally, the monster weapon opened up, lending its thumping, devastating firepower to the madness and the mayhem. From the slow, steady beat I guessed it had to be firing at the rate of a dozen or more projectiles a minute.

As they smashed into the earth, a series of rhythmic flashes lit up the Green Zone, throwing monstrous shadows across our path. I felt as if I was running through a tunnel of churning, howling, raging fire, and into the depths of hell itself.

As the awesome firepower chased us up the track, it was touch and go as to whether we'd keep ahead of Spooky's pounding annihilation. One wrong move by the pilot and a lot of us were going to get whacked. I spurred my legs to move faster, and cursed myself for smoking so many cigarettes over the previous two days. I reached the high ground soaked in sweat and coughing my guts up, but still very much alive. The adrenaline was pumping in bucket-loads. I didn't know which had been the more terrifying: the mad dash through the kill zone, or being under the Spooky call sign's guns.

With awe-inspiring skill the warplane's crew had managed to open an escape corridor for us all the way back to the desert, without putting a shot wrong. Not one of the lads on that crazed mission had so much as a scratch on him. It was miraculous. It was like we were blessed. I knew in my heart it couldn't last.

There had been something of *Indiana Jones and the Temple of Doom* about that feint mission into the midnight heart of darkness. But there was also something Monty Python-esque about the crazed rush of the escape: 'Run away! Run away! Run away!'

At 0100 we were back in our desert laager, and I got a call from the Spooky call sign overhead.

'*Widow Seven Nine, Spooky Two Zero*. Task force has extracted, mission accomplished. Thanks for having your boys go in for us like that.'

'It was nothing, mate,' I replied. 'If you're done 'n' dusted, can we go back to sleep now?'

I heard the pilot chuckle. 'Roger that, sir. I'm gonna do a sweep around your position, just to make sure you guys are safe to get some shut-eye. Stand by.'

I was tempted to ask him not to bother, and to let me get some kip, but I knew what these American pilots were like. You couldn't fault their enthusiasm, or their skill. A few minutes later a call on the TACSAT jerked me out of a doze.

'Sweep complete. There's nothing moving in the desert. You get your heads down. And stay safe down there.'

'Same to you lot,' I mumbled.

My head slumped back on to the sand, and I was out like a light.

By 0700 the following morning the SIB and bomb-disposal boys were done. We went about the horrific task of zipping up the nine bloodied corpses into body bags, and loading them on board one of the Vikings. We received the order to mount up. It wasn't a moment too soon to be getting the hell out of there.

I got my head out the wagon's turret, so I could see what was what. I never bothered wearing my body armour in the Vector, as it slowed me down too much. With the weight of the Osprey ceramic plates, I'd never be able to haul my fat arse out of the turret, or do my job properly.

On the signal for the off, there was this massive explosion to the front of us, not more than five metres away. It blew me back inside the Vector. Everything was chaos and confusion, and I didn't have the slightest idea how I was still alive.

Through the ringing in my ears I could hear the distant sound of Chris screaming at Throp to reverse. There were voices everywhere yelling for us to dismount, but they sounded as if they were coming from the end of a long, echoing tunnel. A thick, choking smoke was everywhere. I presumed we'd gone over a mine and that the wagon was on fire. For fifteen seconds or more I was blinded in dust and a toxic burning that had me gagging for breath. Eventually I managed to haul myself back out of the turret, and as I did so I caught sight of what was on fire.

The Vector in front of us was billowing smoke and flames. Guys were falling about half-obscured in the thick cloud of acrid black fumes, as they tried to evacuate through the rear door of the vehicle. One of them lunged for our wagon, but he didn't make it, and collapsed on the dirt right in front of us.

'The Vector in front's been blown to fuck!' I yelled. 'It's the wagon in front!'

'Get the lads into our wagon!' Chris yelled back. 'Get them in here! Now!'

Sticky and I vaulted down and fought our way through the smoke. I grabbed the nearest body and dragged him inside the wagon, then went back for another.

There were six lads from Somme Company, a Territorial Army (TA) unit attached to 2 MERCIAN, in that Vector. By some miracle all were still alive. The worst was a lad with a smashed leg. We manhandled him into the rear of our Vector in double-quick time.

As the smoke cleared it was obvious this was no mine strike. A jagged rent had been torn in the side of the Vector, where some kind of projectile had torn it open like a giant tin opener. Only a direct hit from a 107mm rocket could have done that.

The wagons were on a ridge above the wadi, and the enemy must have targeted us from out of the Green Zone. Our Vector was directly in the line of fire of the next rocket. We got the last guy loaded and the rear door to the Vector slammed shut.

'Throp, fucking step on it!' I yelled. 'Get us the fuck out of here!'

The last words weren't out of my mouth before Throp dropped the clutch and we shot forward. As he flogged the Vector and it bucked and smashed its way across the rough ground, another 107mm fired, smashing into the desert somewhere to the rear of us.

Together with the Somme lads we were crammed into the back like sardines. We held on to the injured soldier to stop him from cannoning off the ceiling with all the bumps, and I got my one free hand on the TACSAT.

'*Widow TOC, Widow Seven Nine,*' I yelled above the noise of the speeding vehicle. 'Sitrep: one Vector blown up, one injured T3. Need immediate CAS.'

'*Widow Seven Nine, Widow TOC.* Roger that. I'm sending you *Hog Two One,* fifteen minutes out.'

CAS stands for Close Air Support – the nearest aircraft that is airborne and can come to a unit's aid. I had a Hog call sign inbound, which meant an A-10 tank buster was on its way. I reckoned the injured lad was no more than a T3 – the least urgent casualty – and that he'd make it back to FOB Price fine in the Vector.

I radioed the OC and gave him a heads-up. My main concern was the burning wagon, which was stuffed full of all sorts of sensitive comms kit, plus crates and crates of mortar rounds. I couldn't believe how none of those mortars had blown up when the 107mm tore into the wagon.

We rejoined the main body of the convoy and waited for the A-10 to pitch up. The lads from Somme Company were going apeshit in the back of the Vector. All they wanted to do was get back to FOB Price, so they could phone their mates and let them know they'd been 'blown up in Afghanistan'. It was Friday night back in the UK and they reckoned we could make it back in time for night-club chucking-out time. They were all painters and decorators and the like, and they wanted to get on to their mates to have a good crow. You had to hand it to them: *that's the spirit, lads.*

The injured soldier was on an adrenaline high. As the 107mm had hit, it had blown a mortar clean out of its crate. He'd watched the round exit via one of the Vector's open hatches and come flying back down through the other, whereupon it had smashed into his thigh. And that was how he had sustained his injury. The mortar hadn't exploded, of course, or else none of the Somme lads would be breathing. But whatever way you looked at it, the fact they were alive and in one piece was a bleeding miracle.

With the A-10 inbound, Butsy decided he and some men had to return to the scene of the attack. They needed to rip out the Vector's most sensitive equipment, in case the A-10 didn't destroy it all, plus there were mortar rounds in there that we couldn't allow to fall into enemy hands. They set off in two vehicles and loaded up the ammo from the burning Vector.

Just as Butsy had finished what was about the craziest mission imaginable, they pulled away from the burning Vector and another 107mm slammed into the desert where they'd been sitting.

The A-10 checked in to my ROZ and I did my easiest talk-on of the deployment so far. We were six hundred metres from the burning Vector, which was throwing a pillar of thick, oily black smoke high into the sky.

'See that column of smoke,' I told the pilot. 'Vehicle to the base of that. Take it out. Your choice ordnance.'

'Roger that,' the pilot chuckled. 'I was visual with that smoke forty nautical miles out. I'm gonna hit it with a Mark 82 five-hundred-pound laser-guided bomb.'

A minute later there was a blinding flash, followed a couple of seconds later by a deafening blast. The damaged Vector was smashed to pieces. BDA was a direct hit, as if we needed it. We watched over that Vector until it was completely burned out, and we knew there was nothing of use the enemy could scavenge from it.

Then we were ordered to make tracks. En route to the base the A-10 was ripped by a pair of Harriers, *Recoil Four One* and *Recoil Four Two*. And under the watchful gaze of the British jets we were shepherded back through the gates of FOB Price.

We took the guy with the injured leg and dropped him at the sickbay. It turned out that he had a bruised – *not broken* – femur, where the mortar round had crunched him. None of the other Somme lads had the slightest injury – not even burst eardrums. Unbelievable.

It was 1400 hours when we gathered as a company for the post-op debrief. None of us had so much as managed a wash or to get any scoff. Major Butt addressed the men.

'This op was a massive success,' he began. 'The ferocity and sophistication of the enemy showed what they can do, but this mission also allowed us to put into practice what we've trained for, and to prove it. We worked in the FST and MFC on the high ground, like we'd trained for, and now it's been tested against a toughened enemy. It was good to see the FST plugging us constant air for the first twenty-four hours.'

The OC paused for a second. 'The mission went on for four days not one, reflecting the ferocity of the contact. The bravery of you lads brought us all together, and built confidence in each other in terms of what we can do. There are lessons to be learned. We need more signallers, to keep HQ informed. We need more medics, so we have one embedded with each platoon.

'We know the sergeant major needs the means and protection to move around the battlefield and do resupply at will. The need for better resupply hindered offensive operations. At the time we had the right equipment in the right amount, but it was touch and go on water and ammunition. Next time, we'll get those things right.'

After the OC's upbeat briefing, we got the Intel assessments. Some fifty-plus enemy fighters were reported killed in the battle for Adin Zai, including four senior commanders, plus one complete mortar team had been taken out. My call sign – *Widow Seven Nine* – had thirty-two confirmed kills, from 6,500 pounds of bombs dropped, plus strafing runs. B Company had just the one injury – the bruised femur resulting from the flying mortar. That was it: a top op.

After the briefing I went to have a much-needed wash. The shower block was a length of canvas tenting, divided into cubicles. Outside each was a container of Army-issue disinfectant, all part of the drive to prevent the spread of nasty bugs. Water was rationed.

You had to push a button, race inside your cubicle, lather up and shower before it stopped flowing.

I'd got myself a new tube of shower gel from the NAAFI. I stood inside my cubicle and let the warm water run over me. It felt like paradise. I lathered up my hair with the shower gel, and started to rinse. But the more I rinsed the more soap kept pouring down my face, and I couldn't get the stuff out of my hair or my eyes. I knew the water was about to run out, and I cursed the bloody NAAFI for selling me some dodgy shower gel.

I tried scrubbing harder, but the more I rubbed the more suds there were. I was just starting to lose it, when from the cubicle beside me someone cracked up laughing. I'd know that laugh anywhere: it was Throp. An instant later I could hear Sticky creased up on the other side, and I realised what the two of them had been up to.

As I'd showered, they'd both been pumping Army-issue hand-cleaner on to my head – the bastards.

'Fuck off, you tits!' I yelled. 'Or I'll knack you!'

Throp and Sticky made themselves scarce, and at last I could scrub my balls in peace.

After, I retired to my tent for some well-earned kip. As I drifted off to sleep I reflected upon how we'd proven ourselves in battle. I have to admit it – I was feeling pretty good about it. But little did I know that we were about to get sent back into Adin Zai.

And this time, the enemy would be waiting for us.

SEVEN
GOING BACK IN

I came to my senses in the pitch dark. Someone was shaking me. It was Sticky, and he was muttering something about me having air. For a few moments I thought I was having a nightmare, for who could have allocated me air at FOB Price?

What do I need air cover for here, I felt like screaming. I'm in FOB Price. LEAVE ME ALONE. But it was hardly Sticky's fault, so no point shooting the messenger.

Someone had booked me in for an Air Special Request from 0400–0800. An Air Special Request (ASR) is air cover booked in advance, for a specific reason. Why the hell someone had given me an ASR at FOB Price I couldn't figure. But I was told that no one else could use it, so I had to.

It was a bit chilly, so I pulled on my battered yellow Hackett jumper and a pair of shorts and flip-flops, and hunched over my TACSAT. I had a giant B-1B supersonic bomber overhead – call sign *Bone Two Three*. All I could think of getting the poor bastard to do was fly recces over the desert, searching for enemy vehicles on weapons resupply missions. What a waste of a monster air asset like a B-1B.

At 0615 I told the pilot that I was going off station for a few minutes, and asked him to continue flying air recces. I couldn't resist the smell of frying grub that was drifting over from the cookhouse. I was starving, and I just had to get a brew and some scoff. I left Sticky to monitor the air, and told him I'd be back in a jiffy.

—

I hurried over to the mess tent, piled a plate high with sausage, egg and bacon, perched on a deserted bench and started to ram it in fast. I felt a presence beside me. I glanced up from my plate to see the camp commandant standing on my shoulder. He was staring at me with ill-disguised contempt.

'What d'you think you're doing in mixed dress?' he snapped.

It was forbidden to wear shorts around the main British Army bases, and I guess the captain hadn't gone a bundle on the jumper, either. If nothing else it obscured any badges of regiment or rank I might've been wearing, so he had no idea who I was.

'I've been up since 0400 controlling a B-1B,' I answered, speaking through a mouthful of grub. 'I'm going right back to it.'

'I don't give a damn,' the captain replied, stiffly. 'Mixed dress isn't acceptable …'

'Right, well, like I said I'm just trying to get some bloody breakfast before going back to me jets.'

I couldn't abide blokes like that. They stalked around in their starched kit and never left the base. There were a few more words between us, before I finished my grub and stomped out. The captain kept firing comments at me even as I left the mess tent.

I knew the camp commandant did a daily briefing to all senior commanders at 0800. It was 0745, and I asked the B-1B pilot if he could fly a low-level show of force over FOB Price, just to deter any enemy that might be eyeing us. I asked him to bring his jet in at 0800 sharp, right over the top of us.

'I can come down to five thousand feet,' the pilot told me. 'Any lower I have to clear it with higher.'

'I don't think that's enough of a deterrent,' I replied.

'What d'ya mean?' the pilot queried.

'Any enemy who's dicking us to attack won't be deterred by that. I need you lower.'

—

The pilot put it up to his superiors, and got cleared down to two thousand feet. I confirmed that was good enough, and repeated that I wanted the show of force at 0800 sharp.

'What you up to?' Sticky asked.

He'd been listening in on my shoulder. I explained to him the squabble I'd had in the cookhouse, and that I was planning to buzz the camp commandant's briefing.

'Don't be daft,' Sticky told me. 'You'll be right in the shit.'

At the same time he couldn't help laughing. Throp was awake by now, and just as soon as he heard what I was up to he started egging me on.

Bang on the nose at 0800 this massive swing-winged aircraft came screaming over the walls. It was like having a jumbo jet land in your back garden, only worse. The B-1B tore FOB Price to shreds in terms of the audio level, and left our ears ringing.

'*Bone Two Three*, excellent show of force,' I radioed the pilot. 'Any enemy watching FOB Price would've definitely been deterred.'

The pilot signed out of my ROZ, and no sooner had he done so than a runner arrived at our tent. I was summoned to see the regimental sergeant major, who accused me of ruining the camp commandant's briefing. He'd heard about our argument in the cookhouse, and he demanded an explanation for what I'd been up to with the B-1B.

'Show of force to deter any enemy dicking FOB Price, sir,' I told him.

'Fuck off. There's been no attacks on FOB Price for months, as you well know.'

I was ordered to go and apologise to the camp commandant. I went to see him, and said I was sorry for being so abrupt in the cookhouse over my mixed dress.

'It won't happen again, sir,' I assured him.

—

73

'Thanks for apologising, Sergeant Bommer,' he replied. 'Doing the manly thing and all that. I never knew who you were, actually.'

He didn't mention the B-1B, so I left it at that. I returned to our tent feeling honour had been satisfied, but Chris collared me and tried to give me another bollocking.

'Bommer, you can't bloody do stuff like that,' he was saying, 'It's just not on. It'll build a barrier between us and them ...'

But at the same time as he was trying to be serious, Chris couldn't help laughing. I headed to the gym with Sticky, to pump some iron and work it out of my system. We made a good training partnership, and whenever we were back at FOB Price we'd hit the weights together.

I ran into Butsy in the gym. He gave me a thumbs-up, before taking me aside for a fatherly chat. 'If you ever have problems in the cookhouse or whatever, Bommer ... I know what you did and why you did it, but you're one of mine, so come and speak to me about it. I won't tolerate people speaking to my lads out of turn like that bloke did you.'

I knew I'd got a gentle bollocking, but I knew why. And what Butsy had said just served to reinforce the sense that he considered all of us in the FST as his own.

Once I was done in the gym I managed to place a phone call home. My wife, Nicola, had seen a lot on the news about lads getting killed and injured in Helmand, and she was understandably worried.

'You are OK, aren't you?' she asked me. 'You would tell me?'

' 'Course I would,' I reassured her. 'Nowt's been happening our end,' I lied. 'I've not heard a thing about anyone getting injured or anything. It's nowhere near here and nowt to do with us lot, anyhow. So come on, tell me: how're the nippers?'

Nicola told me that Harry and Ella, my infant son and daughter, were just fine, but missing their dad, which I liked to hear.

'Roger-dodger. Well, put 'em on the line then.'

—

I had a couple of words with my little ones, by which time my ration of phone minutes was almost done. The Army allowed us twenty minutes' talk time a week. Nicola came back on the line.

'So, have you been using your new job?' By that she meant JTAC-ing.

'Yeah, a bit,' I told her.

'So how is it?'

'Tell you the truth, it's class.'

Nicola and I had been married for seven years, and she's a fine lass. She'd had a good education, as her dad was in the Army and he'd served all over the world. The one thing he'd been dead against was his daughter marrying a soldier. So when Sergeant Grahame came along I was hardly flavour of the month. Things were all good now, mind, and Nicola's dad and I were spot on. But I had no illusions as to how tough it was on a young wife with two kids when her man went off to war.

After finishing JTAC school I'd been sent for six months of continuous training and exercises, to get me combat-ready. I'd been all over the UK, Canada and the US, doing drops with dummy bombs and then live ones out on the ranges. In that six months I'd had two weeks off at Christmas with the family. Then I'd had my final JTAC exams, passed, and that was it – I was off to Afghanistan.

I'd first gone to a real war – as opposed to a ghost war like Northern Ireland or Bosnia – in 2003. The night I was leaving for Iraq I'd gone into my son's bedroom to say goodbye. It was 3am and he was fast asleep. I went to give him a kiss, and suddenly I was all choked up. I couldn't stop crying. Christ, I thought, this could be the last time I ever see him.

By the time I got to The Light Dragoons' base, I was still all red-eyed.

'What's up with you, mate?' one of the lads had asked.

'Nothing. I've just been saying goodbye to the nipper. I ended up blubbering.'

Well, that was it – a load of the other lads came right out and admitted the same. They'd been crying too, when saying goodbye to their women and their kids. They'd been standing outside the gates for twenty minutes chain-smoking fags to try to dry their eyes out.

It wasn't easy on relationships, being away fighting a war. But as far as I was concerned that's what I'd signed up for at age seventeen, when I'd joined up. All I'd ever wanted was join the British Army. Since the age of four when I was given my first Action Man soldier, soldiering was all I'd ever been interested in.

At first I'd done pretty well at school. I'd got nine O-levels, and my parents were chuffed as nuts with me. Trouble was, school just wasn't for me. I was the only one in my class with tattoos, and with the burning desire to be a soldier. I left school and got a job with a local locksmith, to kill time until I was seventeen and could sign up. I did a lot of work assisting the police breaking into cars that had been stolen and abandoned. That was pretty handy – learning how to nick cars at age sixteen.

But the owner of the locksmiths was having trouble with his marriage, and one day he came into work drunk and tried to take it out on me. Being a punchy lad I floored him and was given the sack. I was on the cusp of my seventeenth birthday, so I told my dad I was off to join the Army. He told me that I'd never last two weeks. 'Two weeks and you'll be out,' was what he said. My dad's words were like a red rag to a bull.

The Light Dragoons are based in the north-east of England, so lads from my neck of the woods automatically went into them. When I went to sign up the recruiting officer told me that the regiment operated behind enemy lines, in tanks. I'd always wanted to drive a tank, so that was me in.

I lasted the first two weeks of basic training, and after that there was no looking back. I joined The Light Dragoons on 14 February 1996. On my first day one of the lads, a bloke called Gary 'Baldy' Wilkinson, nicknamed me 'Bommer'. It was pretty obvious how Baldy had got his nickname: he had alopecia, a disease that makes the hair fall out. But I didn't have a clue why he'd decided to call me 'Bommer'.

When I asked, he said he'd named me after Herol 'Bomber' Graham, a black British boxer. His career had peaked in the late 1980s, but he was trying to make a comeback. After that the name just stuck. A year later and if you asked the lads for Paul Grahame, they'd not have a clue who you meant. Everyone knew me as plain 'Bommer'.

The Light Dragoons are a Formation Recce regiment. Their role is to forge ahead of the main battle group in small units, gathering intelligence on the enemy. Troop positions, areas of special interest, high-value targets – those were the kind of elements that interested us. In formal Army speak, our role was to enable a 'done-by-recce-pull' – pulling the main battle group through, with its big tanks, convoys and troop numbers.

But that was conventional war fighting, and originally designed to combat a Soviet threat. In Iraq and Afghanistan we were waging a totally different kind of warfare. We were up against insurgents who wore no uniforms and did their best to hide amongst the local population. In Afghanistan in particular, The Light Dragoons formation recce concept had to be radically redrawn.

In Helmand the new soldiering ethic was to work as small, highly mobile units independent of resupply for days at a time. We'd carry all of our food, water, fuel and ammo with us, using CRVT (Combat Recce Vehicle – Tracked) vehicles for cargo-carrying and mobility. We'd be a recce and strike force, with sniper teams, Javelin missile units, and Scimitar light tanks providing firepower.

The role of the JTAC was central to this new concept of war fighting. Working behind enemy lines, we'd have eyes and ears prior to other units, placing JTACs in an ideal position to smash any targets of opportunity. The JTAC could call in airstrikes where the unit didn't have the firepower, or the reach, to hit. That's how I'd ended up being put up for training as a Light Dragoons JTAC. By then I'd been ten years in the Army, and I was a qualified crew commander, which meant I'd been trained how to command and fight my own Scimitar light tank. It was rock-hard to get on to the JTAC training, and I was dead happy to be put up for it.

Prior to Afghanistan, a lot of soldiers had trained as JTACs, but they'd never really got to use their specialist skills. Even in Iraq, commanders had failed to use the JTACs properly, or to integrate them into battle plans. Few understood the JTAC's capabilities or role – that of being integrated with the fighting troops, and calling in danger-close air missions on the front line.

But my course was specifically tailored to Afghanistan, and there was a feeling that in Helmand, the JTACs were really going to come into their own. I started at JFACTSU (Joint Forward Air Control Training and Standards Unit), based at RAF Leeming in north Yorkshire. JFACTSU has a winged tommy gun and pair of rockets as a cap badge.

My instructor was a Corporal Grant 'Cuff' Cuthbertson, and he'd been out in Helmand serving with the Gurkhas, and doing the job for real. He told me that as a JTAC, I'd get to see action for sure in Afghanistan. But first, there was the best part of a year's training ahead of me, for which Cuff would be my mentor and guide.

Being a JTAC, the instructors explained, was about bringing the biggest and the best weapons systems to any party with pinpoint precision and accuracy. It made perfect sense to me. It was the mechanics of it that were so challenging. We started with the fundamentals – learning the theory of bringing in low-, medium- and high-level air attacks, and what munitions to choose for which target.

Then we moved on to map-reading, and how to plot targets. My dad was big into hill walking, so my compass and orienteering skills weren't all bad. I'd pretty much mastered them in the Army. But working to latitude and longitude grid references was all new to me, and a real mind game. We were tested every week, on the dreaded Friday afternoons. Fail those Friday tests, and the instructors would bin you.

You could pass all the exams, but that still didn't mean you'd make it as a JTAC. It was the instructors' role to ensure we trainees had those certain, intangible qualities that were required of a fully combat-ready JTAC: you had to have the gift of the gab, and to be able to think on your feet, whilst splitting your mind into many different dimensions all at once.

This was how the instructors put it: *imagine you're doing a low-level attack at fifty feet, and you have sixty seconds to get the pilot's eyes on target and let him have time to arm his weapons systems and release. You're doing the talk-on, and at low level like this there's no second chance; he'll not do a re-attack, for the risk of being shot down is too great. If you can't think and talk that fast you won't hack it as a JTAC.*

They taught us big, medium and small – as the priority of features to talk the pilot on to. You'd start big, choosing a distinctive woodstrip, white building or a yellow field between two green ones. Once you'd got the pilot visual with that, you'd go smaller, and finally to pinpoint detail. With each talk-on you had to 'see' it all from the pilot's perspective. You had to put yourself in the cockpit, and imagine his view of the battlefield over the nose cone of the aircraft. From there you'd use the clock-distance-object method for the talk-on. For example: *three o'clock from the parked vehicle, at one hundred metres' distance, bunker position.*

That aspect – doing the talk-on from the pilot's perspective – was what most of the trainees failed on. JFACTSU is purely an intellectual course, with no physical aspect to it. With most of the lads having come from elite units, it was taken as read that we were up for the physical side of things.

There were twelve on my course, and at the end I passed out 'limited combat-ready'. I was given a certificate, a team photograph of the lads on the course, plus a black leather-bound logbook, for entering my controls. Months of dry-run training and exercises would follow before I could take my new skills to war.

Cuff coached me over the months ahead, becoming my Sub-FAC, or mentor. He watched over me in the UK as I did dummy runs on the ranges, using aircraft dropping concrete bombs. Then in Canada and the US he 'daddied' me, as I did my first controls using jets dropping live ordnance and firing live rounds. By the time I was done, Cuff had become a true mate. I was his creation, his JTAC, and he'd moulded me in his image. When I passed out, there were 178 combat-ready JTACs in the entire British military. There were precious few of us to go round.

With JTACs being in such short supply, I'd been immediately posted to 2 MERCIAN for the six months of their Afghan tour. 2 MERCIAN were a tough infantry regiment, whose soldiering ethos relied on taking the fight to the enemy on foot and at close quarters. I couldn't have asked for a better bunch of lads to be fighting with.

And we'd need to be at the absolute top our game for the next mission. We'd be going deep into the Green Zone, where we'd not only be taking enemy territory, but holding it.

Few if any British soldiers had occupied territory this deep in the Green Zone. We were going into the unknown.

EIGHT
BUM LOVE TUESDAY

A couple of days after my low-level stunt with the B-1B at FOB Price, Chris returned from a briefing with the OC, and told me the good news. Intel reports had the enemy reoccupying their positions at Adin Zai. They'd reinforced the place with hundreds of fighters from their base further east, at Siurakay. They were making a stand at Adin Zai, and we were going back in to take them on.

This time round, the OC's plan of attack was markedly different. At first his orders were to take the company back into the Green Zone on a full-frontal assault, but he had refused to do that. The enemy had learnt well the lessons of our first battle. Butsy was convinced they would allow us to advance, then surround us at close quarters so we couldn't use the air.

Instead, he planned to strike first at the village of Rahim Kalay, to the east of Adin Zai. From there we'd hook back round, taking Adin Zai from the opposite direction and by surprise. He demanded and was given full air cover for the entire duration of the mission.

He also had an elite Czech Army unit placed under his command. Butsy was well pleased: we'd worked with the Czechs before; they were crack soldiers and they'd never let us down. They were a force multiplier par excellence. He'd used the Czechs to harass the enemy from the flanks, as the company went in to attack.

There was one other crucial difference in the coming battle: once we'd taken the ground, we were to hold it. There would be no pulling out. We were to take enemy territory deep in the Green

Zone, and make it our own. Adin Zai – and Rahim Kalay with it – was going to be B Company's stalking ground for the remainder of our tour.

We began the mission to retake Adin Zai by throwing out probing patrols, to test the enemy's forward lines. We were returning in convoy from one such patrol, when we got a radio message: I was to be dropped at Patrol Base North (PB North).

Two fortified patrol bases (PBs) had been constructed a kilometre back from Adin Zai, on the desert high ground. Intel reports had the enemy massing for an attack on one of those bases.

It had been a bad day already. Earlier I'd got news that we'd lost a Danish JTAC. I'd never met *Norseman Two Two*. He was working further up the valley, towards Sangin. But we'd chatted over the air, and I felt like we were good mates. His vehicle had gone over a mine whilst he was doing a live drop, and he'd been killed outright.

The news that I was to be left that night at PB North made it a total shit of a day. PB North was occupied by eight Afghan National Army (ANA) soldiers, none of whom spoke more than a few words of English. Some British Army lads were supposed to be joining us, but it was 1300 and there was no sign of them. I wasn't the slightest bit happy.

There was no sense in leaving a JTAC alone in a base occupied by soldiers with whom he couldn't communicate. Orders were that I had to stay to coordinate the air, but if that was the case why didn't I have a couple of lads with me? Why didn't Sticky or Throp get to stay? I'd feel a whole lot better that way.

It was a load of bollocks dumping a lone British squaddie in a base populated solely by Afghan soldiers, for there were also plenty of tales about how bastard bent they could be. But the order that I was to stay had come direct from the top, for the patrol bases needed a JTAC. As our convoy pulled away from the base leaving me behind, I felt pissed as hell.

I glanced around PB North – the place where the lads had abandoned me. It was a triangular structure build from massive HESCO Bastion walling – rectangular, wire-framed cages filled with rock and earth. On each corner of the base was a sangar, a fortified firing point. Just one, gated entrance led into the base.

It wasn't so much the fortifications I mistrusted, it was the guys that I'd been left here with. No one was saying a great deal, for we didn't have the words to communicate, but the Afghan soldiers did offer me some local unleavened bread. As I took a token nibble, they asked me what I did as a soldier. I pointed at the bread and mimed kneading a loaf. They were staring at me like I was totally cracked.

'Chef,' I told them. 'Army chef. Cooking. I bake bread and make scoff for the lads.'

I reckoned if I told them I was a JTAC, that would only give them a bigger incentive to sell me out to the enemy. I was paranoid I was going to end up with a new uniform – an orange boiler suit, like those poor bloody hostages in Iraq.

I'd noticed that there were two Toyota four-wheel drives parked up in the base. I wondered whether I should nick one and do a Mad Max and drive the half-mile down to PB South, for I knew there was a British Army contingent down there.

Before I could make a decision either way, the Afghan soldiers mounted up, two to each truck, and without a word they roared out of the gates. That was that, then. I wouldn't be stealing a Toyota any time soon. Maybe they were heading off to have a chat with their Taliban mates, to ask how much a fat British chef was worth.

It was a sticky, sweaty, burning hot afternoon and I was in a foul temper. I'd spotted a well in a deserted compound to one side of the base. I decided to go and have a wash and a cool down. I grabbed my rifle, my spare mags and my JTAC kit, and wandered over.

I was stripped naked and tipping the cool water over my head, when I felt a presence behind me. I didn't give a fuck that I had my three-card trick out: I spun around, making an instinctive lunge for my SA80. Standing there were two soldiers from the base. They'd obviously been having a good look at me stark bollock naked, and lathered in soap.

I noticed that they were holding hands. I also noticed that they were both wearing bright red nail varnish. No British soldier who'd spent more than a few days in Helmand could fail to have heard rumours about our Afghan colleagues' reputation for man-love. Apparently some of them were about as straight as a bloody roundabout.

We'd all of us been told about 'Bum Love Tuesday' – or was it Thursday? – the day of the week when they'd get it on with each other. With my rifle levelled and a few choice gestures, I made it clear that wasn't my bag. They were to fuck off and leave me alone.

They turned and left for the main base. But I was boiling now. I was fuming. As I rinsed off I just kept thinking: *How the fuck can they leave me in the middle of bandit country alone with a load of ANA I can't even communicate with! And what's the point in bloody washing myself? I'm only getting clean so I can put on my orange boiler suit when the kidnap team return.*

I finished washing, returned to the base and got on my TACSAT – the only means I had of communicating with FOB Price. I dialled up Damo Martin, the captain who ran Fire Control Planning (FCP) cell – the JTAC's ops room back at FOB Price. Damo was a fellow JTAC and a down-to-earth kind of bloke. I'd never been one to mince my words, and I gave it to him straight.

'*Widow Eight Two*, this is *Widow Seven Nine*: I want to know what the fuck is going on!' I fumed. 'What the fuck're you lot doing, leaving me up here? *On my own?* What a totally fucking twatful situation to leave me in …'

No one could explain why I'd been dumped at the base on my own, without even an interpreter. It was all a mix-up of communications and planning. Butsy was massively pissed off that I'd been left

Aerial view of the Helmand River and the heavily-
vegetated Green Zone to either side of it. This is the
terrain we would fight in during our six-month tour.

Battle stations. The rooftop position at Alpha Xray, the British base set in the heart of
enemy territory, on the banks of the Helmand River. The tripod-mounted .50 calibre
heavy machine gun, GPMGs, Javelin missiles and other hardware testify to the ferocity
of the fight, as the enemy sent wave after wave of fighters to assault and overrun our
position. I brought in repeated danger close air strikes to smash the enemy all around
Alpha Xray, beating off their attacks from the very walls of the base.

Everything – even your brains – melted in the intense Afghan heat. The only way to eat a Yorkie bar was to rip off one corner and suck it all down. 'Not for Civvies' indeed.

Me in the Vector 'borrowed' from Camp Bastion. I'm on my TACSAT, there's the Rover Terminal propped up on some ammo boxes, the map opposite and GPS hung from the ceiling.

Calling in a 1000-pound Joint Direct Attack Munition (JDAM) from a Harrier. You can just make it out in the top left of the picture plummeting earthwards before hitting the Sangin compound.

With a ten millisecond delay the 1000-pound JDAM penetrates the roof and the first floor of the concrete building, and then detonates.

Moments later the JDAM unleashes its full destructive power. The dust and debris cloud towers hundreds of metres into the air. I'm 150 metres from the explosion, having called it in danger close.

A patrol of 2 MERCIAN soldiers sets out from PB Sandford for Alpha Xray, in the heart of enemy territory, driving quad bikes and Snatch Land Rovers.

© Andrew Parsons

Bad hair and beard day. Our Fire Support Team after
months under siege in The Triangle. From left to right:
Chris, me, Sticky, Jess, and big Throp with a Dushka gun.

Handing out the goodies to the Afghan kids.
It was a security nightmare, but the best way
to get through to the locals.

Firing at night from the rooftop of Alpha Xray, and from the WMIK's 50-cal heavy machinegun parked up below, as seen through a night vision monocle.

An AH64 Apache helicopter gunship banks hard over PB Sandford, as it goes into action over The Triangle. With its 30mm cannon, Hellfire missiles and CRV7 rockets, plus unrivaled optics and loiter time, this was my platform of choice in Helmand, and the one airframe that the enemy were known to fear above all others.

The old boy's compound at Monkey One Echo, complete with
Arabian-Knights-like tower, doughnut archway and donkey.

JTAC Central. My favourite position atop the roof at PB Sandford.
It offered all-around view over enemy positions, but zero cover.
That's me with the water bottle, and Jess perched atop the
sandbags, and the OC, Butsy, with his foot on the camonetting.

Returning fire with a mortar barrage from PB Sandford.

A 2 MERCIAN soldier fights off an enemy assault, putting down fire with his Minimi Squad Assault Weapon (SAW) light machine gun, following an ambush at close quarters. The enemy was only prevented from overrunning us when I called in a series of airstrikes right on top of our own positions.

here, and that he'd lost his JTAC. The bad news was that no one was able to come and fetch me today, so I'd have to man it out. A patrol would come to relieve me first thing in the morning.

I bedded down under a sheet of cam-netting that I'd rigged up against one wall. I positioned myself to one side of the main gate, where I could keep good watch on it. I had my TACSAT ready and I did a weapons check.

I had six full mags on my chest-webbing, and one in the SA80 – so that made 210 rounds in all. That should keep me going for a while. I carried a Browning 9mm pistol as back-up weapon, with three, thirteen-round mags. Plus I had two fragmentation grenades, one in a chest pouch and one in my backpack.

I kept an eye on the sangars, and the Afghans did have a sentry in each. Trouble was they kept strolling between the towers, to have a natter. Whenever they did so it left one side of the triangular fort unguarded. I resigned myself to a long and restless night, for no way was I going to sleep in this place.

At last light the Toyotas returned. It seemed they'd been on a bread-run, and not selling me out to the Taliban. The fresh bread actually smelled gorgeous, so I went and got some scoff. But there was about as much crack as a one-inch whip, for we couldn't talk.

Once I'd stuffed myself I got under my cam-net with my loaded rifle, and resumed my position and state of watchfulness. The attack came at 2100.

From out of nowhere a barrage of fire hit the sangar on the south-east corner. I was on my feet and up to the top like a rat up a drain-pipe. Two of the ANA soldiers came pelting after me, so we were three Afghans and me in the sangar, facing however many were hitting us.

'Taliban try break in!' one of the ANA lads yelled.

I flipped my night-vision monocle down over my right eye, pulled my rifle into the aim, and got on the TACSAT. I'd managed to scrounge a TACSAT headset from the Danish contingent at FOB

Price. It clipped on to my helmet, and I could speak into it hands-free, so I could talk and shoot at the same time.

'*Widow TOC, Widow Seven Nine*, d'you copy?' I yelled, as I squeezed off my first round.

Two of the attackers were in the bush fifty metres below, and one of them was hefting an RPG. Muzzle flashes sparked eerily in the green fuzz of the night-vision, the tracer rounds smudging lime-coloured slugs across the distance between us. I watched my own tracer chew into the dirt just short of the RPG-gunner's head, and ratcheted up my aim.

'*Widow Seven Nine, Widow TOC*, you're loud and clear.'

'Sitrep: TIC PB North,' I yelled. 'Need immediate CAS.'

'Roger that. *Rammit Six One* inbound your position six minutes.'

I had a Dutch F-16 six minutes out. In the meantime I was going to have to try to stay alive and kill some of the bastard enemy. Rounds were slamming into the sangar thick and fast. An RPG hammered through the darkness just above our heads.

I adjusted my aim on the RPG-gunner who was darting about like a madman, but as I went to fire one of the ANA lads tried to shove me out the way.

'No one shoot you today!' he yelled. 'I die! I die!'

He tried to get between me and the enemy bullets, as he loosed off what seemed like an entire AK-47 mag with barely an attempt to aim. Nine out of ten for showmanship, but nowt for killing the enemy.

'This is how it's done, you nugget,' I growled. 'Watch my tracer!'

I steadied my aim, got the needle sight on the RPG-gunner's head, and fired. I saw a round plough into the guy's face, and an instant later he was laid out flat and unmoving. All the ANA lads were blatting away, so any one of us could have dropped the RPG-gunner. But in my head I claimed the kill, for their sense of aim was haywire.

As the enemy's gunner's mucker tried to drag the bloodied body away, another leapt forwards to claim the RPG. I readjusted my aim, and fired again and again. The noise on the sangar was deafening, but mostly it was the ANA boys loosing off wild bursts. They had no night-vision, but that didn't prevent them from aiming at the enemy muzzle flashes. Generally, if you aimed just below and to the right of the flash, you'd hit the gunman in the chest or even the heart.

'Listen, fucking calm down!' I yelled at the ANA lads. 'Do as I do. Nice, aimed single shots, that's the way to nail 'em.'

I had a horrible feeling it was going to be a long night's fighting. As we traded fire with the enemy, I gave the ANA boys an impromptu lesson on fire discipline, and how to conserve their ammo.

We hadn't killed that many when the contact died down. I didn't know if the enemy had called it off, or if they were regrouping to try another line of attack. I didn't like the situation one little bit. There were nine of 'us' and god only knows how many of them.

Had I been with eight 2 MERCIAN lads, I'd have happily defended that base all night long. As it was, this was the worst moment of my entire tour. I had dark visions of the enemy battering down the gates – or being let in by some traitorous ANA bastard – and of capture and torture or worse. Well, no way was that going to happen. I vowed to keep a last bullet for myself. If it came to it, I'd put the last round in my head and blow my brains all to hell.

I got *Rammit Six One* flying recces all around the base. For forty-five minutes the pilot didn't spot a thing. Finally, I got a call from Widow TOC that the jet was coming off station, as nothing was happening. I persuaded the pilot to do a final low-level show of force right over the top of us.

By now the ANA were staring at me, wondering who the hell I was talking to on the TACSAT. I guess they suspected I wasn't your average chef. They could hear the jet whizzing about overhead, but I had no idea if they'd connected it to me or not. I hoped they hadn't. I'd much rather remain a cook in their eyes. I asked the pilot to make his run from east to west, so he came in right across the

Green Zone. And I asked him to achieve sonic boom as he passed over us, just to wake the ANA up a little.

They were gabbling away nineteen-to-the-dozen about the attack, when the jet appeared over the eastern sangar, almost kissing the top of the HESCO. It loomed out of the pitch black like some monster alien spacecraft, and pulled up into a screaming climb, spitting cartwheels of fire.

The ANA lads were in a right flap. The three sentries abandoned their sangars, and I watched in utter amazement as one of the lads sprinted out of the gate, haring off into the Green Zone.

So there I was in PB North with no sentries, seven Afghan soldiers cowering in the centre of the fort, and one AWOL somewhere in the Green Zone. Smart. I sat up straight, back to the wall, and cradled my gun. I was in my position beneath the camo-net, and I wasn't moving or sleeping. If anyone came through the gate with hostile intent, I'd kill them. I laid out some tabs before me, and sparked up the first. I inhaled deeply, and settled down for a long night.

I had a lot of time to think. I reflected upon the guy that I'd just killed. It was unusual for a JTAC to see your round tearing into the face of the enemy. Normally, you'd be dropping the big munitions at a distance, not killing up close and personal. The death of that RPG-gunner had been graphic, and the image was burned into my mind.

I asked myself if it bothered me that I'd whacked that guy? Killing that enemy fighter didn't really worry me; it was either him or me. I'd first knowingly killed a man in Iraq, back in 2003, at the height of that war. I was out in Al-Amarah with a patrol from The Light Dragoons running a Vehicle Check-Point (VCP).

It was 0300, and I was the gunner on the Scimitar light tank. One of the lads, a softly spoken South African, called Rob Deery, stopped a car. He asked the driver to open his boot, but the driver refused. He asked again. Again the driver refused.

I jumped down from the wagon and faced up to the Iraqi. 'Open your fucking boot.'

The guy gave me the evil eye. 'No open.'

I forced him to hand over the keys. I cracked open the boot and it was crammed full of AK-47s. I told the guy if he came to the base and produced permits for the weapons, we'd return them to him. For now, we were confiscating the lot. The guy drove off, and a couple of minutes later we were engaged from three hundred metres away. The fire was coming from the Al-Amarah football stadium, where the Iraqi police had a checkpoint. It was clear as day that it was the police who were engaging us.

I dropped to the deck and returned fire with my SA80. One of the lads, John Hunter, asked me what the hell I was doing. I told him I was winning the bloody firefight.

'But they're Iraqi coppers,' he objected.

The Scimitar was acting like a bullet magnet, and that's where the rounds were hitting. I ran around the back dodging fire and jumped into the gunner's seat. As I did so, I cracked my head on the turret. That was it: I was steaming now.

We headed out in the Scimitar to find the fuckers who were shooting at us. As we approached the football stadium, I spotted two guys on the roof. Both had AK-47s. They opened up just as soon as they saw us coming. I was perched on the backrest of the gunner's chair with my head sticking out of the wagon's open turret, rounds ricocheting all around me.

I dropped down, grabbed the Gimpy and returned fire. I hit the first guy in the left eye socket and then in the heart. As he fell through the air his weapon was still spewing out tracer rounds. The second guy was in cover behind a horizontal concrete railing. I put the first round in his right shin, two more in the left, and four across the belly – so that was below and above the pillar I'd hit him.

When we got back to base and dismounted the vehicle, I saw all the impact marks of the bullets that they'd been spraying at us. I'd had my 'tankie' headset on during the contact, so I hadn't heard the racket all the incoming rounds had made. They were bunched all around where my head had been poking out of the wagon, and just inches away.

It was either those guys or me, and luckily I'd got the drop on the both of them. But the story that got back to Nicola was that I'd taken out a whole village of insurgents with a Leatherman. I got on the phone and told her the truth – that two guys had tried to take us out with AK-47s, whilst we were in a tank. I still had a bit of explaining to do though. Prior to deploying I'd told her that we were going out 'litter picking' for the Paras, so there was no way we'd be getting into any trouble.

Yet there was a crucial difference between almost getting shot in Iraq and almost getting shot in Afghanistan. During the early stages of the Iraq conflict, we knew who the enemy were, and we could see who was shooting at us. In Afghanistan, you'd get engaged and not have a clue where the shot had come from.

It was partly due to the terrain in the Green Zone, but increasingly down to the sneaky-beaky way the enemy operated. Being shot at and not being able to shoot back was a real pain in the arse. And it felt very good to have slotted that RPG-gunner, and to have beaten back their attack.

At 0200 the ANA guy who had done a runner returned to the base. He had to knock on the front gate to get let in. He didn't even have his weapon on him. It was mind-boggling. He'd bolted into the hostile darkness without even his gun. His mates bawled him out for being a useless bloody soldier, I guessed, although I couldn't understand a word of what was said.

For twenty minutes after he'd legged it no one had been manning the sangars, as the sentries were too scared to go up again. I guess they feared I'd pull another stunt with a jet. During that time the fort was unguarded, and I expected the enemy in through the gates at any moment. I knew I shouldn't have brought that F-16 in so low: it was about thirty feet above the walls. But I just couldn't resist it. The ANA lot probably knew by now that's what I did for a living. Maybe that guy had run off into the Green Zone to tell his Taliban mates: *We have a super-JTAC in the base – get the orange jump suit ready!*

I knew that JTACs got put where the threat was greatest – that was our role. It's what we trained for.

It was being left here alone that I couldn't stomach.

NINE
DREAMING OF A MAGNERS ON ICE

I woke at 0530 slumped against the HESCO wall. The last thing I could remember was smoking a tab at 0400, and wondering when the fucking sun would get a move on. I must've dropped off.

There was no stand-to with the ANA, so I got on the TACSAT to FOB Price. The patrol had left at first light, so they should be with me any time soon. An hour later there was the growl of engines, and a convoy of vehicles pulled to a halt outside the gates.

I had never been so relieved to see a bunch of British squaddies in all my life. It wasn't Sticky, Throp and Chris and my FST wagon, but frankly I'd have been happy to see Dad's Army coming over the hill. It looked like I wouldn't be swopping my desert combats for an orange jump suit any time soon. Top job.

The patrol consisted of twenty 2 MERCIAN lads, under the command of Lieutenant Greg McLeod, an absolute monster of a bloke. He made Throp look positively weedy. Greg might have been monstrous, but you couldn't have met a nicer bloke. He didn't have a bad bone in his body. He was so nice I reckoned I'd have fancied him, had I been the red-nail-varnish-wearing type. Greg and I sat on some empty wooden pallets, as his lads took the piss out of the situation that I'd been left in. *A lone Tommy with a couple of mags of ammo and his Afghan bum-pals* – it was loads of that kind of thing.

By the time the lads had finished ripping the piss, pretty much all of my anger was gone. You had to see the funny side. That was the thing with the British Army: your average soldier dealt with the very worst of situations by constantly taking the piss.

The patrol wasn't going anywhere, so I took the chance to get some much-needed kip. I woke a couple of hours later to a mouth-ful of choking dirt. We were in the midst of a howling sandstorm, and muggins here had been sleeping with his mouth open. As I coughed and spat, Greg had a good laugh.

'Bommer, mate, can you imagine getting your hands on a Magners on ice,' he remarked, in his booming voice. He mimed pouring the bottle into a glass and taking a long pull. 'Imagine it, mate. Now. In your fist. A chilled Magners.'

The triangular-shaped base had no shade whatsoever. All there was to drink was bottled water that had been half boiled in the heat. From that moment onwards I became obsessed with the idea of a Magners on ice. Not a day went by when I didn't think of it, plus I dreamt of it at night.

Once the sandstorm was over I asked Greg if I could use his satphone to phone the wife. Each platoon carried its own satphone, as part of its 'lost comms' procedure. If radio contact went down, the satphone was pretty much a bulletproof back-up. We could use our Army phone cards on the satphones, and Greg was more than happy to oblige.

It was a big day for Nicola – her twenty-fifth birthday. Before leaving FOB Price I'd managed to order her flowers, chocolates and balloons, all via the internet. You had to love the wonders of modern technology, and I was dying to find out if she'd got them. I went and found me a quiet corner and dialled the number. As soon as the call went through I started to sing her 'Happy Birthday', and we both ended up laughing.

'You are good to me,' she said. 'I got everything – the flowers are lovely! I'm chuffed to bits. Amazing what you can do on the internet, even when you're out in Afghanistan!'

'Well, you just enjoy yourself, birthday girl.'

'So, what've you been doing?' she asked. 'Been up to anything new?'

I just told her it was the same-old same-old. But then added, 'Tell you what, do me a favour and buy a crate-load of Magners cider, and stack the fridge full of it. A crate, mind. I want one in a pint glass full of ice, just as soon as I get home.'

Nicola said she'd get on to it. I came off the phone feeling dead happy. It was amazing what a call home could do for your spirits. A few hours back I'd been convinced I was going to end up kidnapped or dead. Now, all was good with the world. Family: it's crucial, as far as I'm concerned.

I always kept five minutes of my weekly phone card ration to phone my mum and dad. And every other day I'd get a letter off to Nicola and the nippers. Nicola teased me that they were full of spelling mistakes. I'd written those letters in the heat and dust of an Afghan desert in the middle of a war, yet she had the neck to moan about the spelling.

Over lunch of some boiled ratpacks Greg briefed me on the company's plans. The assault to retake Adin Zai was scheduled for two days' time. My FST would pick me up en route to the attack. Greg's platoon would remain with me at PB North, to deter any attacks on the base.

The following morning I was ordered to move down to PB South, which Intel suggested was now the enemy target. The British soldiers who were located there came to fetch me in a couple of WMIKs. They were part of the 'omelette' – Operational Mentoring and Liaison Teams – programme, tasked to train our Afghan allies.

I was wounded to be leaving the 2 MERCIAN lads, until I realised that PB South was right on the banks of the Helmand River.

I got a short briefing on the base – which was a carbon copy of PB North – ditched all my kit, and jumped in the river fully clothed. It was deep and the current was fast, but it was fantastic to cool off and give the uniform a good scrubbing.

That night I got allocated some air. I had a French Mirage, call sign *Simca Three One*, flying air recces over the darkened terrain. I was up in one of the sangars chatting to the pilot, when I noticed a couple of furtive figures below. They were moving along one wall in the direction of the hydro shed – a tin shack that housed a hydro-electric generator.

I got my night-vision on to them, and it turned out to be two of our Afghan soldiers. I watched them disappear into the hydro shed. What they were up to, skulking around in the dark? Were they planting a bomb or something? I got the night-vision on to the one window, but at first I couldn't make out a thing in there.

Finally I realised that I was spying on a couple of Afghans who'd gone for a good hump in the privacy of the tin shed. I started hurling rocks, but after several, clanging hits, I realised that there was no stopping them. So be it, I thought. Each to their own.

I opened my peepers at 0250 and got my kit packed away. The omelette boys gave me a lift to PB North, where the 2 MERCIAN lads were just stirring. We had two platoons laagered up in the desert, just short of Rahim Kalay, and two platoons here at PB North. Everyone was wired for the coming assault, and no one had slept much.

I had one ear on the TACSAT, monitoring the air and seeing if any warplanes became available. If they did, I wanted them over our lads in the desert as a protective umbrella. Whilst doing so I got chatting to a Sergeant Mikey Wallis, a Royal Artillery bloke attached to 2 MERCIAN. He asked me what I did and I explained that I was the JTAC. I asked him the same, and he told me he was the LCMR Operator.

'What's an LCMR?' I asked.

'Lightweight Counter-Mortar Radar,' he told me. 'Basically, I locate mortars when the enemy fires them. I've got a radar-like gizmo that does the business. Gets it narrowed down to a ten-figure grid.'

Mikey sounded like a pretty useful bloke to know. Prior to now he'd been stuck in Camp Bastion, and we'd got him allocated to us because of the mission to retake Adin Zai. He had a TACSAT, so we swapped frequencies and call signs.

'This is how we'll work it,' I suggested. 'If you get a mortar signal, give me a call to warn the lads. And pass me the ten-figure grid. Then I'll take a look from the air.'

Mikey nodded. 'Will do, mate.'

'There'll be a lot of traffic on my frequency between me and the jets,' I added. 'But if it's a mortar grid it's crucial, so just cut in.'

'Right-oh,' he confirmed. 'I'll be doing my stuff from PB North. I'll keep you posted from here, mate.'

At 0500 the FST wagon pitched up. Throp, Chris and Sticky gave me a chorus of how're you doing, Bommer, mate? The piss-taking bastards.

'I'm all right,' I replied. 'Great time I've been having with me bum-pals here on me tod.'

I dumped all my kit beside the Vector, and the four of us started talking through the coming mission. The brief was to remove the enemy from Rahim Kalay and then Adin Zai, pushing them further east into the Green Zone. As with the previous assault, we would be on the high ground overlooking the battlefield.

'There's Intel coming down that the enemy know the "tank" on the high ground controls the air,' Chris added. 'And that it's calling in the bombs.'

'That'll be us, then,' I remarked.

'They know what we're doing,' Chris continued, 'so they've more than likely planted mines on the high ground.'

'Anyone know if the wagon's mine-proof?' asked Sticky.

All four of us kind of shrugged

'Only one way to find out,' Throp grunted. 'Drive over one.'

'I'd prefer it if you didn't,' said Sticky.

'Aye, me an' all,' I said.

And that was it – we were good to go. We were 1.5 clicks from the line of departure for the assault. We had to be up on the high ground before 0700, zero hour for the attack. The lads had identified a small re-entrant (a kind of cutting) on the ridge where we could position ourselves in overwatch, but still have a little cover.

With Throp helping me, I went to strap my Bergen on the outside of the Vector. There was little room in the back, what with all our gear. It was around 0640, and just as I was attaching the pack there was a series of massive explosions over towards the Green Zone. A firefight had kicked off somewhere in the direction of where we were heading.

I left Throp holding my pack, and dived in the rear of the wagon. I didn't give a damn about that Bergen any more. I wanted air cover like yesterday.

'*Widow TOC, Widow Seven Nine*. Sitrep: troops in contact, request immediate CAS.'

'*Widow Seven Nine, Widow TOC*. Wait out.'

It was still thirty minutes away from my first ASR (Air Special Request) that I had booked for the mission. Widow TOC had to be checking what platforms they had available and in the air.

'Move out,' said Chris. 'We need to get to the demarcation line asap, to get eyes on the contact. Fuck the mines, if there are any!'

Throp wrung the Vector's neck, and we bounced and cannoned our way into the open desert. En route we got a sitrep from the OC. The enemy must have clocked the 2 MERCIAN lads as they moved in towards Rahim Kalay. By the time our platoons had reached the line of departure, they were well ready. From out of nowhere the enemy had opened up with a savage mortar barrage.

In the time it took our wagon to hare its way across the desert, they'd got the fifth round in the air and zeroed in on our positions in the Green Zone. It was mayhem.

The lads were shit scared of those mortars. At the same time they were being hit by automatic weapons and sniper fire, and RPG rounds. The OC made it clear that he wanted us up on the high ground, so we could start smashing the enemy from the air.

Two minutes after setting out from PB North we pulled up on the ridge line. Luckily, Throp hadn't driven over any mines – or not ones that had exploded, any road. I stuck myself out of the Vector's turret to get eyes on the battlefield. As I did so, the first thing I noticed was the howl of an incoming mortar. Throp hadn't even managed to find a parking space when the round smacked into the desert a hundred metres beyond us. It wasn't bad for a first shot. From the howling of the mortar rounds there had to be more than one tube in action. And with the firefight raging right below us the battle noise was deafening.

'*Widow TOC, Widow Seven Nine*, where's my air?' I yelled into the TACSAT.

'*Widow Seven Nine, Widow TOC. Dude One Five* and *Dude One Six* in your overhead in eight minutes.'

I had a pair of F-15s inbound. The F-15 can achieve two and a half times the speed of sound at altitude. That was how I was getting the jets overhead only ten minutes after they'd been scrambled.

I put a call through to the jets. '*Dude One Five, Widow Seven Nine*, do you copy?'

'*Widow Seven Nine*, this is *Dude One Five*, I have you loud and clear. Inbound your position seven, repeat seven minutes. Standard loads and ninety minutes' playtime.'

'Standard loads' meant a regular ordnance package for an F-15, and I had them overhead for ninety minutes. That should be more than enough bombs and time to knock seven bales of shit out of the enemy. Now we just had to find them.

'Sitrep: I have three platoons in the Green Zone at grid …'

BOOM! The last words were lost in the roar of an explosion as a mortar ploughed into the dirt not twenty metres short of us. Time to get moving. As Throp reversed like a lunatic across the barren terrain, Sticky was holding on to me to stop me flying out of the Vector's turret. I tried to continue briefing the F-15s.

'Repeat – friendly grids are 98057238 …'

'Break! Break!' came a voice on the net. '*Widow Seven Nine, Nine One Charlie*. I have a grid for you of that mortar that just fired: 3748567389. Repeat: 3748567389. It's firing from the south side of the Helmand River.'

'Roger, enemy mortar grid is: 3748567389,' I repeated to Mikey.

Mikey Wallis and his mortar-tracking gizmo had come up trumps. It wasn't a moment too soon. The enemy mortars were bracketing Butsy and his HQ element. The OC and his lads were having to sprint for a new patch of cover every third round that slammed into the dirt, or else they were going to get splatted. Mortars were smashing into the bush twenty metres from them, and Butsy was complaining that trying to command whilst eating dirt wasn't very easy. There was nothing those poor bastards down in the GZ could do about the enemy mortar teams, for they were well out of their range. Only we could hit them.

He came up on the net, yelling above the deafening crack and thump of battle: 'Bommer – you need to sort those mortars! Like now!'

'Roger. *Dude One Five, Widow Seven Nine*, I want you to search for a hot mortar tube at grid: 3748567389. Readback.'

The pilot confirmed the tasking and the grid. No sooner had he done so than there was a faint boom in the distance, and another mortar came howling down. This one slammed into the Green Zone just metres from our troops.

There was another boom, and a second shell tore into the thick bush around our lads. A third went up, this one tearing into the rock and sand where the Vector had just been sitting.

'*Widow Seven Nine, Charlie One Zero*,' came Mikey's voice on my TACSAT. 'I have two more grids for you. Repeat, two more enemy mortar grids.'

Mikey passed me the coordinates. We now had three enemy mortar teams in action. One was targeting us lot, whilst the other two were dropping rounds on top of the lads below. It was complete carnage, and the platoons hadn't even crossed their line of departure.

With two fast jets to control, Mikey muscling in on the net, three enemy grids to plot plus our friendlies, I had my hands full. I left Sticky and Chris to liaise with the OC, whilst I concentrated on finding and smashing the enemy. The OC had told us to crack on and get the bloody job done, and we knew he had every confidence in our abilities.

I decided to split the aircraft. '*Dude One Five*, I want you searching for enemy forces around our forward line of troops. *Dude One Six*, I want you overhead those three mortar grids, to find 'em and smash 'em.'

'*Dude One Five*, affirmative.'

'*Dude One Six*, roger that, sir.'

A tense few seconds followed as the F-15s began their searches, their sniper optics scanning the terrain below. *Dude One Six* was the first to come back to me.

'I got a PID on three males around a straight heat source, three metres from the last ten-figure grid you gave me. It's 2.7 kilometres from your nearest friendlies.'

'Roger. Wait out.'

Before I'd said a word Chris was clearing it with the OC. 'OC says he's pinned down as are all platoons,' Chris yelled over to me. 'They need fucking space to move out from under those mortars ...'

BOOOOOM! Another mortar round slammed into the dirt fifteen metres from the wagon. It rocked the Vector like a ship caught in a hurricane, shrapnel and rocks pounding into the wagon's steel sides. No doubt about it, the enemy mortar operators were bloody good. They were close and getting closer.

'*Dude One Six, Widow Seven Nine*,' I yelled, above the echoing noise of the explosion. 'Attack from whatever line is fastest, your choice ordnance!'

'Roger. I'm banking up now to do a vertical dive on to target.' A pause. 'Tipping in and requesting clearance.'

'You got clearance,' I yelled. 'Clear hot! Proceed with the attack!'

Bugger the procedures – we and the lads were getting smashed. Plus I needed my frequency clear to talk to *Dude One Six*'s wing.

'*Dude One Five, Widow Seven Nine* – I need a sitrep.'

'Sitrep: visual four armed pax with RPGs and small arms three hundred metres from the lead element of your troops. Visual …'

The pilot's last words were lost in an enormous crack, as whatever ordnance it was that *Dude One Six* had dropped slammed into the earth on the opposite side of the river. Get in! Hopefully, that was one less bastard mortar team to deal with.

'*Dude One Six*, BDA,' I asked, as a mushroom cloud of smoke billowed above the impact point.

'BDA: two pax killed, mortar tube is fucked. But one pax fled, ten metres from bomb impact point and got away.'

I was about to retask the pilot to join his wing searching for enemy fighters in around our platoons. But instead he had this for me.

'I'm watching four armed males run away from that first mortar grid you gave me. I'm visual with them going into a tiny mud hut, two metres by two metres, more like a garden shed.'

'Confirm the four pax are armed, and no civvies are in the target vicinity,' I asked.

'Affirmative. I have PID'd them with weapons, and I am happy under rules of engagement to proceed to attack.'

'Roger,' I confirmed. 'Wait out.'

'Chris!' I yelled. 'Get the OC. Do we hit 'em or what?'

Chris got clearance from the OC, and I passed it up to the pilot. 'Confirm enemy pax are still at the target.'

'Affirmative. There are no other doors to the building, and I've been watching it like a hawk. Sir, no one's come in or out of there.'

'Right, I want a GBU-12 dropped on target using a south-to-north attacking run.'

'Affirmative. Tipping in. Call for clearance.'

'*Dude One Six*, you're clear hot.'

'In hot,' came the pilot's reply. Then, 'Stores.'

There was a thirty-second delay as the arrow-shaped munition streaked through the air. I hoped and prayed that the enemy mortar team didn't decide to leave their garden shed. There was a stupendous crack as the eight-hundred-pound bomb hit, throwing up a dense cloud of dust and smoke on the horizon to the east of us.

'*Dude One Six*, BDA,' I requested.

'BDA: mud hut and occupants obliterated. They're all dead, sir.'

I figured, that was two mortar teams taken out, plus one mortar tube. There was one team left to hunt for, but it looked as if they'd been warned that I was smashing their buddies from the air. The blokes were still in contact, but no more mortars were going up now. Plus we'd just received some intercepts suggesting the enemy knew exactly what the lads and I were up to.

'Look for the man with the stubby black antenna,' a Taliban commander kept yelling over his radio. 'He controls the aeroplanes. Target him, and the tank on the high ground. The tank speaks to the jets that are hitting our brothers.'

There was nothing we could do about it. I got the F-15s flying low-level shows of force over our forward positions, as they probed for the enemy. *Dude One Five* came back on the air to me.

'*Widow Seven Nine*, I'm visual with four male pax at the second, ten-figure grid you gave us. They're gathered around a tube that I'm sure is a mortar.'

Result! We'd found the third mortar team. Chris radioed the OC, and the message came back to attack.

I felt a burst of adrenaline. '*Dude One Five, Widow Seven Nine*. I'm clearing you in. Ordnance and attack line of your choosing.'

'Affirmative. Tipping in.'

I watched the dart-shaped form of the F-15 banking around in a fast but graceful turn, its twin vertical tail fins slicing through the sky, the gaping intakes to either side of the fuselage sucking in the air.

'*Widow Seven Nine, Dude One Five*,' came the American pilot's voice. 'Sir, we got a problem. I'm visual four kids in the vicinity of the mortar tube. And sir, one of the mortar team is holding a kid right beside the tube.'

Shit! The pilot was tipping in and I had seconds to make the call. What the fuck did I do? If I cleared the bomb, I was as good as murdering four innocent children. If I aborted, the mortar team would send up more rounds to smash our lads.

'*Dude One Five, Widow Seven Nine*, wait out.*'

I yelled down the hatch into the Vector's interior, 'Chris, the fucking pilot's got four kids at the mortar tube! I can't fucking do it! That bastard will play tricks with my head for the rest of my life …'

I didn't bother completing the sentence. '*Dude One Five, Widow Seven Nine*. What d'you reckon?' I asked the pilot.

'*Widow Seven Nine*, it's you who buys the bomb,' the pilot replied.

'Then I can't fucking do it. I got two kids at home. I can't do it. Abort! Abort!'

'Affirmative: aborting the attack. And many thanks for that call, *Widow Seven Nine*. I wouldn't have done the run for you anyway. I got kids back home.'

Barely moments after I had aborted that airstrike, the call came up on the net that every soldier dreads.

'Man down! Man down! MAAAN DOWWNN!'

The instant we had that 'man down' call it all went horribly quiet on the net.

It was the first 'man down' call we'd had of the deployment, and no one could quite believe it. For several seconds the entire company seemed to hold its breath, and the jets I was controlling went completely out of my mind.

A voice broke into the silence. '*Charlie Charlie One*, roger, go firm.' It was the OC. As always, he was right in the thick of it. 'All stations: win the firefight. Orders two minutes.'

The OC's words unleashed all my pent-up emotion. I felt the red mist of animal aggression rising. There were enemy fighters out there using kids as human shields, whilst our lads were getting smashed. But I had to hold my anger in check, or I'd lose the ability to do my job properly. All I knew at this stage was that we had a badly injured lad somewhere down in the Green Zone. And right now, we had to get him the hell out of there.

TEN
MAN DOWN

'*Widow TOC, Widow Seven Nine*,' I yelled into my TACSAT. 'We've got a man down! Repeat: man down! We need immediate IRT.'

IRT was the Incident Response Team, a Chinook with medics and an Apache escort on permanent standby at Camp Bastion. They sat on the flight line 24/7 waiting for emergencies like this one. It was the JTAC's role to get the IRT in the air.

'Roger that. What's the severity of the injured?' the Widow controller asked me.

'No position to tell you,' I snapped back. 'But I need IRT right now.'

By the time the platoon commander had got me the casualty report, with the soldier's ZAP number – his unique British Army ID – I knew this lad was in a very bad way. He was classed as a T1, the severest casualty level possible.

Time seemed to slow to a crawl as I waited for the IRT to launch. The horrific thought crossed my mind that it might have been a mortar round that had taken our man down. If so, had my saving those four Afghan kids' lives resulted in one of our lads getting smashed?

Major Butt had always told us that if we had a man down, the focus of the company would immediately switch to extracting the casualty and getting the lads out of the shit. In the carnage that was going on all around us the net went berserk. *Win the firefight.* Those had been the OC's words. Everyone knew that we'd lost someone, and the fire from our side was targeted now with a burning anger. The lads were using accurate shots to put the enemy down.

'*Charlie Charlie One*, all stations,' Butsy's voice came on the net again. 'Orders: 6 Platoon, extract with casualty. 5 Platoon, move to river to give covering fire. 4 Platoon, secure river crossing. Sergeant major to recce route back to LZ. Somme Platoon to provide rear security and secure LZ. FST no change.'

It was a kick-arse set of orders. Under heavy fire the 2 MERCIAN lads had pushed across the river that lay to the north of the Green Zone. To extract the casualty they'd have to cross back over, and Butsy's focus was on securing that river crossing. The Landing Zone (LZ) for the Chinook was set in the open desert halfway to PB North, and the lads from Somme Company would secure it.

The Czech unit were to stay on the high ground, hitting the enemy's northern flank. They were driving Toyota jeeps, complete with DShKs – pronounced 'Dushkas' – a monster piece of kit. The Dushka is a Soviet-era 12.7mm anti-aircraft gun. It can fire only in automatic mode, putting down six hundred rounds a minute. Those rounds can chew their way through walls and trees, and hopefully they were doing just that to the enemy positions right now.

It was my job to get the Chinook into that LZ, plus I still had my jets to control. At the same time I had at least one active mortar team, and I couldn't bring the helicopter in with that still firing. One mortar down on the Chinook, and we would be in a world of pain. There was a squelch of static, and I grabbed the TACSAT.

'*Widow Seven Nine*, *Widow TOC*. *Ugly Five Three* is bringing in the heavy call sign. Expect IRT to be with you in two-five, repeated two-five minutes.'

'Roger that,' I replied. The casevac Chinook was twenty-five minutes away.

I put a call through to the F-15s. '*Dude One Six*, *Widow Seven Nine*. Sitrep: we have a man down and platoons are extracting. I want you to fly repeated shows of force over the enemy positions. If you spot any enemy fighters, you're to smash 'em.'

'Affirmative,' came the US pilot's reply. 'Commencing shows of force now.'

'*Dude One Five, Widow Seven Nine*. I want you overhead that mortar grid, looking mean and nasty. If there's a moment when those fuckers aren't holding kids around the tube, I want you to smash 'em.'

'Affirmative. They won't be gettin' any second chances, *Widow Seven Nine*.'

It took an agonising sixty minutes for 6 Platoon to fight its way to the borders of the Green Zone. At times the lads were crawling along ditches carrying the casualty, under intense sniper fire. At others they were chest-deep in the river, passing the wounded man from shoulder to shoulder as machine-gun rounds whipped and snarled overhead. In the process, two more lads were wounded.

For the last thirty minutes I'd been arguing fiercely with the Chinook pilots to remain on station orbiting over the desert. They were running low on fuel and getting anxious, but we were desperate to get our wounded men out.

'We're nearly there!' I kept telling the pilots. 'We're nearly there!'

Finally, with the Chinook sipping air, the 2 MERCIAN's sergeant major, a real champion of a bloke called Jason 'Peachy' Peach, decided some drastic action was required to get the wounded blokes out. He was in a WMIK and volunteered to go in and get them. Along with Corporal Hill, his driver, and one of the medics, he set off from the high ground into the Green Zone.

The trouble was, a sharp ravine bisected the ridge line, and it lay between their position and the wounded. The only way to skirt round it was for Peachy to drive into the Green Zone, passing in front of the entire company and heading into the enemy guns. The lads were still taking massive fire, and as soon as Peachy's WMIK pitched up in the jungle it became the focus of the enemy attack.

As rounds slammed into the vehicle and RPGs roared overhead, Peachy and the medic blatted away with the WMIK's 50-cal and Gimpy machine guns. Crashing over ruts and with Corporal Hill driving the race of his life, the open-topped Land Rover somehow made it through without being blown up or anyone being killed. The wounded were loaded aboard, and now Peachy and his lads had to return the way they'd come.

The enemy knew it. They'd set a series of RPG ambushes on the route, and as the WMIK thundered back along the track, with Peachy and the medic trying desperately to keep the wounded aboard, the bush erupted in a wall of fire. At the same time the entire company was pumping rounds into the enemy positions, with the careering WMIK sandwiched in between.

Unbelievably, the vehicle made it back to the high ground, and although it was peppered with bullet holes and shrapnel, not a man aboard had been hit. As the WMIK belted up to the makeshift LZ, I banked up the F-15s to 15,000 feet, to deconflict the air, and cleared the Chinook in to land. The casualties were run up the helicopter's rear ramp and loaded aboard.

In a storm of dust the giant, twin-rotor machine clawed its way into the air, and turned towards Camp Bastion. The casualties were on their way, but by now we knew for sure that we'd lost one. Corporal Paul 'Sandy' Sandford, a nineteen-year-old 2 MERCIAN lad and a real character in 6 Platoon, had been shot by an enemy sniper. Most likely, we'd lost Sandy long before the lads had battled their way through the Green Zone to evacuate him. There is a 'Golden Hour' – the sixty minutes in which every casualty is supposed to be air-evacuated to the Camp Bastion field hospital. In Sandy's case, no matter how quickly we'd got him out we could not have saved him.

Butsy had sent a clearance patrol back into the area where Sandy was hit, to retrieve his body armour and kit, but it was gone. The enemy knew we had a man down, and they would have seen the Chinook go in to pick up the casualties. For the first time in the battle for Adin Zai they had their heads up, whilst we were feeling like a crock of shit.

The entire company was out of the Green Zone, and the contact had died down to just about nothing. There was still the odd RPG and sniper round coming our way, but that was about it. It was 1045, and we'd been fighting for four hours solid, and we were back where we'd started. Things weren't going as planned.

It was at this moment that I got the call that there was a fourth casualty needing evacuating – only this time it was one of them. We had an injured enemy fighter in our custody. He'd been shot twice by one of our lads, and he was an urgent T1. We'd just had Sandy shot in the head and a lot of us wanted nothing more than to slot him, but we knew we were better than that. We'd give that wounded enemy fighter the same relief as we would our own. We'd get him on to a Chinook, and back to the field hospital at Camp Bastion – in spite of knowing that if the enemy captured any of us, we'd face a slow and agonising death. It was all about doing the right thing on a rough day.

Before I could dial up a casevac, the Vector was hit by a savage barrage of 107mm rockets. They must have had more than one 107mm launcher in action, for the warheads came in thick and fast, smashing into the dirt all around the wagon. They were trying to drive us off the ridge line, but the only way we were leaving would be in body bags.

With the platoons gone firm on the edge of the Green Zone, we were the only part of the company with eyes directly on the enemy. They had their heads up and they were dangerous. I still had that pair of jets trying to sniff out their positions, but not a sign of the enemy could be found. They'd just been shooting up our lads big time, yet they'd disappeared into thin air.

Throp got the Vector moving, and we shunted back and forth on the ridge line as the 107mm warheads tore into the burning white of the desert to either side of us. As I clung on in the rear of the sweltering, bucking wagon, there was a squelch of static. I grabbed the TACSAT.

'*Widow Seven Nine, Ugly Five Three.*' It was the Apache pilot. 'Understand you have an urgent casualty. Suggest we reland the heavy at LZ and pick up your T1.'

'*Ugly Five Three, Widow Seven Nine.* Appreciate your offer, but right now we're under a barrage of 107mm fire. It's too dangerous to bring in the heavy. Wait out.'

The 107mm rocket launcher has an 8.5 kilometre range. I didn't know where the bastards were firing from, but the LZ was only a kilometre back from us. A direct hit from one of those twenty-kilo rockets wouldn't do a Chinook any good at all, not to mention the aircrew and the wounded lads it was carrying.

I tasked the F-15s to come in low and noisy, searching for those launchers. I briefed the Ugly call sign to do likewise. And in the resulting lull in the rocket barrage we brought the Chinook back in, and the enemy casualty was loaded aboard. As the Chinook thundered off into the burning skies to the west, I got a call from *Dude One Five*.

'I'm ten minutes past my dangerously low fuel level,' the F-15 pilot informed me. '*Widow Seven Niner*, I'm sippin' on air. I got to bug out right now.'

His wing aircraft was bugging out with him, which would leave us with no air. Just as soon as the F-15s had left the airspace, the 107mms started smacking into the high ground all around us. We were back in contact, and the lads had their heads hung low and were at their most vulnerable.

I got on the TACSAT and demanded air. I got a call from Damo Martin, at the Fire Planning Cell (FPC) cell, back at FOB Price.

'*Widow Seven Nine, Widow Eight Two*. I've been listening in on the air, and I've got you a pair of Recoil call signs inbound, five minutes out.'

'Roger that,' I replied. 'And thanks, mate. We fucking need 'em.'

Damo was a class act. He'd been monitoring my frequency, and even before he'd heard the F-15s were leaving he'd dialled up the Harriers. He knew I was up to my eyes in shit. He'd just anticipated what I needed and got it done. It was a top job.

I called up the Harrier pilots. '*Recoil Four One, Widow Seven Nine*, do you copy?'

'*Widow Seven Nine, Recoil Four One*, I have you loud and clear,' came back the crisp English accent of the pilot. 'We're inbound your position, four minutes, standard loads. You've got us for four hours on yo-yo.'

'*Recoil Four One*, that's class. Sitrep: I've got three platoons gone firm on the high ground above Adin Zai, having withdrawn from a heavy contact. We lost three lads …'

As I briefed the Harrier pilots, I realised how dire was our situation. We were four and a half hours into the battle and we hadn't broken in to enemy territory. We'd lost the momentum, and everyone had to be wondering if the mission was going to get sacked. I could only imagine how Major Butt was feeling, over with the main body of the troops. The OC was in a hard place. Taking Rahim Kalay was supposed to be the easy part of a mission, a prelude to Adin Zai. Intel had Rahim Kalay slated as a bog-standard Afghan village. Yet we'd stumbled into a bloody hornets' nest, and had been badly stung. 6 Platoon – Sandy's lot – had been well shot up, and they were lucky not to have lost more.

I sparked up a tab and waited for the Harriers. Down at the desert muster, Major Butt was stealing a few quiet 'Condor moments' himself. He could tell the men of 6 Platoon were badly shaken. They were a strong fighting unit, but they were quiet now, and choking back their tears. None of the lads were in a time or place where they could allow themselves to grieve.

Major Butt called up the commanding officer of 2 MERCIAN, and briefed him on the situation. It was 1120 hours and his men were low on ammo and water. They'd lost three, and had gained no territory. The word from the CO was that the mission had to proceed. No matter what, they had to take Rahim Kalay and Adin Zai.

When the OC gathered his platoon commanders together to brief them, he faced one of the toughest moments of his entire career. He told the men they were going back in. He told them they had to get back on this bike and ride it again. And he assured them that he would be there with his HQ element, at the vanguard of the fighting.

Once he'd finished the briefing, the OC came up on the air.

'*Charlie Charlie One*, all stations. Orders: 5 Platoon, 4 Platoon, continue assault as planned. Advance in contact to retake terrain and compounds. 6 Platoon, remain in reserve. Sergeant Major: resupply of ammo and water as required. Somme Platoon to provide rear security. Czech unit to harass enemy as two platoons move forwards. FST, no change. Resupply and rearm as necessary. Zero hour – 1145.'

As soon as I heard that message I had visions of First World War soldiers going over the top again. We were going back in, and we had twenty-five minutes in which to get ourselves battle-ready. And I knew just what we needed: we needed Apaches. The lads were going to have to fight their way back into the same terrain, and it was going to be up-close and brutal. Only by using Apache gunships would I be able to do the beyond danger-close airstrikes that would be required.

I dialled up Widow TOC and told them that no matter what, they had to send us gunships. They told me I'd have two Ugly call signs overhead in twenty minutes. I was getting my Apaches.

The call from the Harrier pilot came just minutes after the lads had restarted their advance. All was quiet in the Green Zone, but my instinct was screaming danger at me. The silence was ominous and menacing, and it set my skin crawling. We were being watched, and I sensed we were being lured into a trap.

'*Widow Seven Nine, Recoil Four One,*' came the Harrier pilot's call. 'I'm visual three males of fighting age hiding something under blankets in a compound to the fore of your troops. They keep looking at the wall in the direction of your advance.'

'Roger that, but what are they looking at?' I demanded. 'Are they looking *through* the wall at our lads?'

'Negative, they're looking at the wall,' the pilot repeated.

'And the bloody bundles ...'

'Contact! Contact!' Sticky started yelling. '5 Platoon's being hit by RPG and small arms from a compound sixty-five metres east of their positions.'

The death-rattle of the small arms and crump of the RPGs exploding was deafening. In an instant I'd forgotten the Harrier pilot's men-who-stare-at-walls, and I was on the TACSAT to Damo Martin. Sixty-five metres was beyond danger-close for missiles or bombs, and the Harrier carries no cannon. I needed bloody Apache.

'*Widow Eight Two, Widow Seven Nine.* We're in contact, and it's beyond danger-close! I need those fucking Ugly call signs now!'

At the moment I finished the call there was a massive explosion, as a 107mm slammed into the ridge line just metres to the north of us. The blast blew me and Sticky off the roof of the Vector, and in through the wagon's turrets. At the same time the noise of battle ramped up in volume, as the chuntering of heavy machine guns added to the racket.

Just then I got the call that I was longing for. '*Widow Seven Nine,* this is *Ugly Five Zero,* do you copy?'

I clambered back out of the wagon's turret. I had the TACSAT jammed against my ear, in an effort to block out the battle noise.

'*Ugly Five Zero, Widow Seven Nine,* go ahead,' I yelled.

'Two Ugly call signs inbound your position ten minutes, standard loads, two hours' playtime.'

'Roger that. Sitrep: I have two platoons in the Green Zone, both under danger-close contact from small arms, machine-gun fire and RPGs. We've got 107mm rockets targeting us on the high ground ...'

I talked the Apache pilots around the battlefield, and asked them to search in the compounds to the forward line of our troops. The Harrier pilots had spotted males of fighting age in those buildings, but they'd yet to kill a single one. To be frank, I was getting well pissed off with them.

I'd just finished briefing the Apaches, when I had a Harrier pilot on the air.

'Near the compound to the forward line of your troops I'm visual with a stationary white saloon car. It looks suspicious.'

'Does it have a fucking weapon on top of it?' I demanded.

'Negative. No weapons or pax visible.'

'Well, it's not fucking suspicious then is it?'

'Well, it's the way that it's not parked under any trees that raises my concern.'

'Wait out,' I snorted.

I didn't bother saying any more. Our lads were getting smashed from four different positions, and the Harriers had still to spot a single enemy fighter. They were flying a £12 million ground attack aircraft armed to the teeth with Paveway laser-guided bombs, yet they hadn't ID'd a single target, apart from an unoccupied white saloon car.

A couple of minutes later there was the distinctive thud-thud-thudding, as rotor blades cut through the air. From the Vector's turret, the squat black forms of the two Apaches were clearly visible powering in towards us. Ugly by name, ugly by nature. Get in!

But before the gunships were overhead, the raging contact died away to zero. The bastard enemy had heard the Apaches coming, and had gone to ground. There was nixy gunfire from anywhere, now that I had my airframe of choice overhead and primed to seek out and destroy. It just went to show how disciplined and professional the enemy could be.

Just as soon as it had gone quiet, 4 and 5 Platoon were up and clearing compounds on the western outskirts of the village. But not an enemy fighter was to be found. It was unbelievable. How was it that one minute they were spraying our lads with gunfire and RPGs, and the next they had gone?

A boatload of enemy fighters couldn't just vanish. How were they doing this?

Where were they?

ELEVEN
WE WERE MORTAL

'All call signs in my ROZ,' I rasped into my TACSAT. 'I want you searching for enemy fighters in the compounds to the fore of our troops. Thirty seconds ago they were malleting our lads from those positions. Find them.'

I got the Harriers and Apaches deconflicted by altitude, with the jets up high, and set them to work. I got another call from *Recoil Four One* about the white saloon car. Apparently, it was obstructing our line of advance through the centre of the village. It was forming a chokepoint, and the pilot reckoned that it might be a massive bomb. Well maybe he had a point, but first I wanted to find and kill some enemy. At 1315, with the lads pushing into the village, I got the call I was least expecting.

'*Widow Seven Nine, Ugly Five Zero*, we have orders to return to Camp Bastion.'

'You are fucking joking me,' I spluttered. 'Tell me you're fucking joking! The only reason we aren't in contact is 'cause we got Apache above us.'

'I've got Higher kicking off big time about aviation fuel. We've been told we've got to leave.'

'Well, you're not fucking going,' I told them. 'I'm not bloody letting you.'

I got on to the OC, and it was crystal clear Butsy shared my sense of anger and abandonment. He and his men were taking a whole world of shit on the ground, with small arms, RPGs and mortars still hitting them. If they tried to advance without Apache, they'd be walking into a series of massive ambushes. Butsy was fuming: *Bommer, get me something else over us.*

I got on the air to Damo Martin. '*Widow Eight Two*, there's no fucking way I am losing those Apache. We've lost three lads already, and it's only Ugly that's keeping the bastard enemy's heads down. I am not losing them.'

'It's out of my hands, mate. The TIC's closed, and those are the rules.'

I knew full well what the rules were. You were supposed to have an active TIC (troops in contact) to have Apache overhead. But as soon as they left us we would have a TIC, so what was the difference?

'Damo, earn your bastard pay grade and tell whoever you need to those Apaches aren't leaving.'

'I can't make that call, mate. It's above my level.'

'Well, get it up to the bastard level that can make that call.'

'I can't authorise it.'

'Then get the bloody colonel to,' I told him. 'He's there with you, isn't he? He wears the crown and a pip. Tell him to keep those bloody Apache over us.'

Damo told me he'd try. I got back on to the pilots.

'*Ugly Five Zero, Widow Seven Nine.* Listen, mate, I'll remind you once and once only: we've had one T4 and two T2s, and the only thing that's preventing more is you being above us.'

'*Widow Seven Nine*, we're low on fuel and we've had comms failure in one aircraft. We have to return to base now.'

'You can't bloody do that!' I yelled. 'It's only you lot keeping the enemy off of our lads.'

The only reply I got was an echoing void of static. I cursed those Apache pilots, yet little did I know that the aircrew had been choking up listening to me. And their comms *had* failed: the 'crypto fill' – the encrypted communications system of one of the helicopters – had dumped. It made the aircraft next to useless, and they had no option but to set a course for Bastion.

As the noise of the Apache's rotor blades faded away on the baking desert air, I got a call from one of the Harriers, *Recoil Four Two.*

'I'm visual with male pax running around the compound to the forward line of your troops,' the pilot reported. 'They're taking up positions at those same walls.'

No sooner had he said it than there was a burst of fire, and a volley of RPGs came streaking towards our positions.

'*Widow TOC* and *Widow Eight Two*, this is *Widow Seven Nine*,' I yelled into my TACSAT. 'We've got a fucking TIC! The Apaches have gone and we're getting smashed. I want those AH back! I want them fucking back at all costs, before someone gets killed.'

'*Widow Eight Two*, fucking right you'll get 'em mate.' It was Damo. 'I'll get a lift to Bastion and launch them myself if I have to.'

'Just get 'em back above me, before more of our lads get whacked.'

'*Widow Seven Nine*, *Recoil Four Two*,' the Harrier pilot cut in. 'Visual male pax carrying around bundles, uncovering and covering them up again, in same compound as before. It looks highly suspicious, but I can't PID any weapons.'

'*Recoil Four Two*, *Widow Seven Nine*: hit that compound. I want you to use a GBU-12 500-pounder on a west-to-east attacking run.'

'Negative. I can't do that. I can't PID any weapons.'

'Well who the hell else would be running around with bundles in the middle of a firefight?' I demanded. 'It's hardly fucking Tom and Jerry, is it? Our forward platoon is getting smashed by fire coming from that very compound, and I want you to hit it.'

'Negative. Under the rules of engagement I can't be certain ...'

I buried the Harrier pilot's words in a string of curses. He may have been technically in the right, but that meant fuck all when we had lads deep in the shit and getting smashed.

Ten minutes later, and with our lads hunkered down under fire, I got a pair of Apaches inbound. At this stage I had no idea if it was the same pilots as before. I started to give a briefing over the air. I mentioned the compound where the Harrier had seen all the males of fighting age, plus the 'suspicious' white vehicle.

My biggest concern was that the enemy were trying to outflank us, after which we'd be well and truly fucked. If they got us surrounded at close quarters, even the Apaches wouldn't be able to help us. Their weapons systems have a small margin of error, and no one wanted to be on the wrong end of an Apache's 30mm cannon fire.

Just as soon as the pair of gunships were audible, the firefight died down to nothing. I could feel myself burning up with frustration.

'Get your lads to go firm in their positions,' the lead Apache pilot told me. 'Go firm until we've completed our air recces. We were overhead thirty minutes ago, so we know the lie of the land pretty well.'

'Roger that,' I confirmed. 'Going firm.'

So, they were the same aircrew as before. I hoped there were no bad feelings about what I'd said to them.

Being an Army Air Corps unit – as opposed to Air Force or Navy – many of the Apache pilots had been soldiers on the ground at one time or another. It meant that they could think like ground troops, with the same kind of instinct. And they could put themselves in the mindset of the enemy, to try to work out where to find and kill them.

At 1445 with the lads still firm, *Ugly Five One* told me he was visual with three males in the same compound as where the Harrier had seen them. They were covering and uncovering 'long bundles', and peering through the walls at our line of troops.

'They're peering *through* the walls, not *at* them?' I queried.

'Affirmative. They've got spyholes in the walls, looking out over your positions.'

At that moment the Harriers told me they had to bug out, for they were low on fuel.

'Keep safe, and watch what you're doing,' the pilots told me.

I snorted. 'No shit.'

With the Harriers gone, I got allocated a lone F-15, call sign *Dude Zero Three*. I wondered if the American pilot would have the same qualms about firing on what had to be enemy fighters as our Harrier pilots had. Somehow, I doubted it.

'*Widow Seven Nine*, *Ugly Five Zero*,' the Apaches were back on the air. 'We're 105 per cent certain those pax in that compound are enemy fighters. What d'you want us to do?'

I got Chris to put a call through to the OC, asking for clearance to fire warning shots. Butsy gave the green light.

'*Ugly Five Zero*, I want you to fire warning shots outside the compound wall, and see how they react.'

'If we put it outside the wall we won't get a reaction,' the Apache pilot replied. 'It needs to be inside the compound – just to make sure they're not doing a double-glazing survey, or something.'

'How big is the compound?' I asked. I was too wired to appreciate the joke.

'Thirty to forty metres square. We can put a ten-round burst of 30mm into the far side of the compound.'

'Roger that. You're cleared to fire.'

The Apache is equipped with a state-of-the-art surveillance pod, which sticks out of the aircraft's nose like an angry zit. It provides unrivalled day and night-vision in close-up detail. It was via that pod that the pilots above me were peering into the compound, the images from their daytime cameras playing on laptop-sized screens in the two-man cockpit.

Each gunship has a single-barrel 30mm cannon slung beneath the forward, gunner's seat. It can be aimed using pistol-grip hand controls, or 'slaved' to follow the pilot's eyeline via a series of sensors mounted in the cockpit. In that mode, wherever the pilot looks and pulls the trigger, the cannon fires. Plus the stub-wings set to either side of the aircraft carry pods of CRV7 70mm rockets, and Hellfire 'tank buster' missiles.

From the chin turret of *Ugly Five Zero* the 30mm cannon barked. A tongue of white flame shot out from the Apache's gun. There was a couple of seconds' delay, and then the heavy-calibre explosive rounds tore into the hard-beaten surface of the compound's interior.

'No change the pax inside that compound,' the Apache pilot reported. 'They glanced up at us; now they're back peering through their spyholes.'

I cleared them to fire a second warning burst. Only one of the figures reacted. He turned away from the peephole, put his arms behind his back as if on a leisurely stroll, and moved down the wall to the next spyhole.

'*Widow Seven Nine*, it looks to me like ambush positions,' the Apache pilot radioed. 'I know how I'd react if someone put ten rounds of 30mm next to me, and it's not like these guys are doing.'

Via Chris I put it up to the OC. Butsy came back saying we were clear to engage. I cleared the Apaches to open fire, and sat back to enjoy the show.

Both aircraft opened up on target. The boom-boom-boom of the 30mm cannons firing was slow enough for the individual shots to be audible to the ear. I counted: 'one, two, three ... ten ... twenty.' Before the last rounds were out of the guns, the first were slamming into the target.

The twenty-round bursts tore into the position, the 30mm shells exploding on and around the western wall, throwing up gouts of mud and shrapnel. Figures came running out of the dust storm, abandoning their positions in their haste to get out of the killing ground.

The Apache pilots tracked the figures as they crossed the open ground and linked up with others, giving me a running commentary as they did so. As the fleeing figures paused, the Apache pilots spotted the weapons that they'd been trying to hide.

The blanket-bundles concealed AK-47s, RPGs, machine guns and shedloads of ammo.

Now the Apaches hit them with a vengeance, 30mm cannon fire tearing up the hard-beaten earth all around the gunmen. Four sprinted for cover, but there was nothing left of the rest. They'd been vaporised, as the heavy-calibre rounds tore into them.

The survivors split up, legging it in all directions. The guns of the Apaches tracked the runners, firing twenty-round bursts that chased them all across the compound, gouts of dirt and shrapnel exploding at their heels. I had to let the Apaches do their work now. I knew they'd brief me as and when they could.

At this point, the lone F-15 came on the air.

'*Widow Seven Nine*, *Dude Zero Three*, inbound to your ROZ. Standard loads, two hours' playtime. Where d'ya want me, sir?'

I gave him an area update. Then: 'I want you flying air recces all around the compound the Apaches are hitting. The enemy's fleeing it like rats leaving a sinking ship. Check the treelines and any other cover. If you see any enemy fighters, smash 'em.'

'Roger that, sir. Commencing my search.'

Chris passed a message to the OC that we'd hit paydirt. The OC ordered the company to go firm as the Uglys did their work. The lads had been in full-on combat for nine hours now, and most had had precious little sleep the night before. They were exhausted, and running on adrenaline. As they took a much-needed breather, Sergeant Major Peach drove a lone WMIK resupply, dumping fresh ammo and water with the platoons.

Sticky and I sat on the roof of the wagon watching the gunships mallet the compound, with repeated attack runs of 30mm. We'd lost some lads and we needed to regain the initiative. There was nothing better to get the blood pumping than seeing a pair of Apaches tearing the enemy to pieces.

'*Widow Seven Nine, Dude Zero Three,*' the F-15 pilot radioed. 'I've been watching the contact and I've seen your Ugly call signs kill fourteen of 'em. Repeat: fourteen enemy fighters confirmed killed.'

'Roger,' I replied.

'*Widow Seven Nine*, Ugly.' The Apache pilot was breathless. 'We have survivors holed up in two dome-roofed buildings to the north of the compound. We're hitting those with Hellfire.'

'Roger,' I confirmed.

From two kilometres out the pair of Apaches lined up on target and fired. Seconds later a paid of black, needle-like objects flashed through the air above the compound, and tore into the roof of the two buildings, hurling up a plume of rock and debris. As the roar of the explosion reverberated around the battlefield and the dust settled, I asked for a BDA.

'Stand by,' the Ugly pilot replied. 'BDA: both buildings direct hit. It's horrific down there. Carnage. It's clearly a big ambush position. There's armed pax running everywhere. The enemy are fleeing into the treeline – engaging!'

The Apache's cannons spat fire again, as they thundered and spun above the compound, raining death from above. They were hitting the 'leakers', the survivors of the missile strike that were fleeing the shattered buildings.

But above the rhythmic thump-thump-thump of the gunship's cannons, there was a new sound now – the staccato roar of machine guns. The enemy fighters were returning fire. Tracer arced and spat skywards, clawing at the Apaches as they hunted in the air.

A fighter broke cover wielding a PKM, a powerful light anti-aircraft weapon. It's a 7.62mm weapon capable of firing 650 rounds a minute and accurate up to 1,000 metres. The gunner sprinted out the compound gate and along the southern wall, keeping to the cover and the shadows.

As he went to open fire, the Apache's cannons roared, and the earth at the gunner's feet erupted in a hail of jagged shrapnel. The dust cleared, and the wounded fighter was seen to crawl, and then fall into a crescent of shadow at the base of the wall. All of a sudden he disappeared.

The Apache pilots zoomed in the cameras in their nose pods. We were about to discover just where the enemy forces had been hiding.

'*Widow Seven Nine*, Ugly. We're visual with an entrance into a tunnel or a cave, at the base of the southern wall of the compound. We're panning our camera along that wall: there are four tunnel entrances, which seem to run beneath the entire compound. Each entrance is half hidden by a pile of straw or hay, or maybe dry poppy stalks.'

'Roger,' I confirmed. 'So the bastards are hiding underground.'

'Affirmative. I can lase the tunnel entrances and pass you the grid?'

'Fantastic.'

The Apache pilot passed me the ten-figure grid of the tunnel entrances, and I scribbled them down in my JTAC log. Now we had an exact fix on where the enemy fighters had been holed up, in between smashing our lads. There was no doubt in anyone's mind now that this was the enemy stronghold. The only things we didn't know were how many of the bastards were in there, what they were armed with, and how exactly they'd been able to hide. Wherever the pilots spotted movement, or the sparking of a muzzle, they slaved the cannon to that flash, and nailed it.

'Visual six more enemy fighters,' the pilot announced. 'Engaging.'

The lead gunship spun on its axis, as it tracked figures sprinting out of the building and making for the cover of the woodland on the southern side of the village. Before them lay a shallow canal, and as they hit it the gunship opened fire. The second gunship opened up from the opposite bank, sandwiching the enemy in a blast of 30mm cannon fire. Gouts of water plumed up like steam, obliterating the enemy fighters.

'*Widow Seven Nine*, Dude.' It was the F-15. 'I got three more dead. Now two more hit. Them Apaches sure are going berserk down there.'

A lone survivor sprinted for the cover of the woodland on the far side of the canal. Both gunships turned their weapons on the tree-line, plastering it with cannon fire. As the 30mm rounds tore into the woods, a storm of shrapnel went ripping through the foliage. Moments later, a series of violent explosions rippled through the shadows beneath the trees.

It looked as if the Ugly call signs had hit the jackpot in there.

TWELVE
APACHE FORCE

'*Widow Seven Nine*, Ugly. Secondary explosions in the woodstrip running along the canal. It looks like a big enemy position. We're lining up for an attack run using CRV7. Are you happy with us using flechette?'

'Chris,' I yelled. 'They're requesting flechette.'

Chris, and Sticky, just stared at me. 'What?'

'It's a CRV7 rocket firing tungsten darts,' I explained. More baffled looks. 'Sticky, put out an all-stations warning for the lads to get their bloody heads down.'

'Ugly, *Widow Seven Nine*,' I got back to the Apache pilot. 'Happy with CRV7 and flechette.'

'Roger. Stand by.'

As *Ugly Five Zero* flew a tight orbit above the enemy compound, searching for new targets, *Ugly Five One* headed out into the desert to the west of us to start his run-up.

The CRV7 rockets are aimed by the trajectory of the aircraft, so the pilot would need to fly down the enemy gun barrels. Each flechette rocket carries eighty needle-sharp tungsten darts – tungsten being one of the hardest metals known to man. It's the stuff they tip bunker-busting bombs with. The pilot would need to get his attack line just right, so as to saturate the woodline, while not nailing any of our lads.

The lone Apache turned and began its attack run. There was a belch of dirty brown smoke from the pods on the stub-wings, as the gunship fired. Four CRV7 rockets streaked away, trailing fire in their wake. An instant later there was a sharp pop as the missiles released their tungsten darts.

The air above the battleground was filled with the ghostly fzzt-fzzt-fzzt-fzzt-fzzt of the projectiles streaking in. I winced. I was no Carol Vordeman, but that was going to rasp. The three hundred and twenty darts struck the target in a hail of devastation, chewing into tree trunks and splintering branches like a massive chainsaw.

'Get in!' I yelled. 'BDA: one-hundred-and-eighty!'

I couldn't resist it, and it sure got a laugh out of Sticky. Even Throp couldn't help grinning.

'*Widow Seven Nine*, Dude.' I had the F-15 on the air. 'I got enemy pax extracting to the south-east of where the Uglys are hitting 'em. I got armed pax going into a bunker position. This is the grid: 98375826.'

'Roger: 98375826. Stand by.'

I glanced at the map, tracing my finger to the coordinates. It was on the far side of the village from our lads. We could hit the bunker no problem.

'Pass the grid to the OC,' I yelled at Chris. '*Dude Zero Three*, I want you to hit that bunker with a GBU-38. Attack line north-east to south-west, to keep the blast away from the Ugly call signs.'

'Affirm target and attack line. But *Widow Seven Nine*, I can do better 'n' that. I'll hit it with two GBU-38s simultaneously?'

'Happy with that.'

'Tipping in. Call for clearance.'

The F-15 began his attack run, as I warned the Apaches of the airstrike going in.

'Clearance,' came the F-15 pilot's voice.

'Clear hot,' I replied.

'In hot.' A beat. 'Stores.'

After several seconds there was an enormous kaboom-kaboom on the far side of the village. A double-headed mushroom cloud of smoke and debris was thrown high into the air, chunks of wall and wood and sandbags spinning off in all directions.

'*Dude Zero Three*: BDA,' I asked the pilot.

'BDA: the bunker's gone. Enemy pax obliterated.'

As the Apaches hunted above the battleground, squirting off ten-round bursts of 30mm at enemy fighters, I passed the F-15 the coordinates of two of the enemy tunnels.

'Dude, I want you to hit those caves with a double-drop GBU-38. I want a bomb in each cave entrance, with a ten-millisecond delay on the fuses to penetrate deep before exploding. Attack run as before. I'm asking Ugly to talk you on to the caves.'

'Affirmative. But I'm running on fumes here, *Widow Seven Nine*, so make it snappy.'

'Roger. Ugly, *Widow Seven Nine*. I want you to talk the Dude call sign on to those caves. You can see it better from the air, and you can lase the bombs on to target.'

'Roger that,' came the Apache pilot's reply. '*Dude Zero Three*, *Ugly Five Zero*. Target is a series of two cave entrances, at the base of the southern wall of the main compound. I'm lasing the first cave entrance now. Confirm you see my laser spot.'

'Visual your spot,' came the F-15 pilot's reply.

'Spot-on: we're lasing the target for you now, Dude.'

'OK, good spot,' the F-15 pilot confirmed. 'I'm starting my attack run now. *Widow Seven Nine*, Dude: call for clearance.'

'Clear hot,' I confirmed.

'In hot.' A pause. 'Stores.'

The F-15 released a pair of GBU-38 smart bombs, programmed to home in on the Apache's laser beam. The 'hot-point' of the laser – the spot where it bounced back from the target – was the lock-on point for the bombs to strike.

A second double concussion rocked the battlefield, as the F-15's thousand pounds of ordnance smashed into the enemy stronghold. The noise of the double blast was muffled, as the five-hundred-pound bombs had burrowed deep before exploding. Each threw up a tight plume of shattered earth and debris, and the strikes looked to be bang on target.

'Dude: BDA,' I requested.

'Stand by,' the pilot replied. 'BDA: both bombs went into the cave entrances. Caves obliterated. *Widow Seven Nine*, I got to go to the tanker, 'cause I'm sippin' on air.'

'Roger,' I confirmed. I radioed the Apaches. 'Ugly, I'm switching foxtrot.' I was changing frequencies. '*Widow TOC*, *Widow Seven Nine*. My Dude call sign is at the refuelling tanker. I need something with a big-hitting potential overhead.'

'Roger, stand by.'

The F-15 could be anything from ten to forty minutes at the tanker, depending on where it was in orbit. I wanted an air platform that could drop bombs. I was also worried that the Apaches had fired a boatload of rounds, and were low on ammo. I got allocated *Recoil Four Three* and *Recoil Four Four*, a pair of Harriers.

As luck would have it the tanker must've been close to my ROZ. I got the F-15 back above me in no time, and I sent the Harriers on their way. Almost immediately, the F-15 pilot spotted more fighters in the main compound. His bombs must have driven them out of the tunnel system. I got him to hit them with another GBU-38, which blasted the compound into shattered heaps of rubble, and there were four confirmed kills.

The pair of Apaches had been in action for a full hour now, smashing the enemy wherever they found them. It was 1410, and this was without doubt the maddest hour I'd spent in theatre. We'd been smashing everything that moved. The company was still firm, but under sporadic fire, and I sensed we were starting to win this battle.

Finally, the Apache pilots turned their attention to that white saloon car.

'*Widow Seven Nine*, Ugly. Not happy with that vehicle. It's blocking the track between two walls, and your lads will have to walk past it to advance into the village.'

'Roger. Stand by.'

All the aircrews had sensed a danger emanating from that vehicle. First, the Harrier pilot – but I'd ignored him. Now, the Apache aircrews. I passed it up to the OC. Butsy said if there was no one in the car we should hit it, to get rid of the thing.

'Ugly, you're cleared to hit that car.'

'Roger. I'll hit it with one times Hellfire, to deny it. Banking around. I'll give you a sixty-second call.'

A few seconds later I got the call from the Apache. The missile fired, a blinding flash of flame yellow blooming on the aircraft's stub-wing. I saw the high-explosive armour-piercing warhead plummeting earthwards. There was a flash of black against the grey-brown of the village, and an instant later the crack of the exploding Hellfire rolled over us.

I was on my TACSAT asking for a BDA, when four further explosions echoed across the battlefield from the same impact point. The blasts must have torn the vehicle to pieces, for I could see chunks of metal spinning into the air to the nearside of the enemy compound. As the explosions died away, a dense column of oily black smoke fingered skywards, and the saloon car was engulfed in a seething mass of flames.

'BDA: direct hit,' the Ugly pilot reported. 'Secondary explosions. It's going up like Blackpool seafront. Looks like it was full of mines or RPGs, or maybe an IED.'

'A big well done, lads,' I radioed the Ugly aircrews. 'Double A-star top fucking job.'

'Happy with that,' the pilot replied. 'We're Winchester ammo and approaching bingo fuel. We need to return to Bastion. The position seems clear of enemy forces, but watch out for the tunnel system running beneath the compound.'

Winchester ammo meant the Apaches had fired off all their 30mm cannon rounds. They were also running short – 'bingo' – on fuel.

'Roger that, Ugly. You guys should know you saved a lot of lives down here today … We lost three, but there'd have been a whole lot more if we hadn't had you above us.'

'We're glad we could come back to help,' the Apache pilot replied. 'We'll ask Bastion to keep a replacement flight on standby, just in case.'

'Roger that. We'll do a BDA by going in on foot. I'll let you blokes know all about it afterwards.'

'Right, we're out of here. Good hunting, *Widow Seven Nine*. Stay safe.'

As the Apaches banked away from the battlefield and set a course for Camp Bastion, I asked the F-15 to do one more attack run. Two of the four caves had been hit: I wanted the third taken out, leaving the one for us to explore on foot. I got *Dude Zero Three* to put a GBU-38 into the tunnel mouth, and then he had to bug out, low on fuel.

The F-15 was ripped by *Recoil Four One* and *Recoil Four Two*. It was 1630 as the Harriers came into the overhead, and both 4 and 5 Platoon were ordered to recommence their advance on foot. As they pushed into the village proper I got the Harriers flying shows of force. The battlefield had fallen absolutely silent. There was no further incoming – not a sausage.

The OC's orders were to get the entire company in to Rahim Kalay by last light. As the foot soldiers advanced, Throp fired up the Vector and we drove in alongside them. There was nothing else moving apart from us lot. We reached the centre of the villages passing by the shredded remains of the white saloon car, which was still a raging inferno.

The OC split the company into smaller units, each tasked to check out enemy positions. The main compound had been totally flattened. It reeked of burning, death and scorched flesh. I'd never known that bone could burn, but it had blazed and vaporised in those airstrikes. There were corpses and bits of unrecognisable, bloody mess everywhere.

Along with Jason 'Peachy' Peach, the B Company sergeant major, I crawled into the one remaining cave entrance. The others had been hit by the GBU-38s and collapsed. By the light cast from our head torches, we could see there were two dead bodies in the far recesses of the cave. Both the enemy dead were still clutching their weapons. I guess they'd crawled in here after being hit by the Apache's cannon fire, and this is where they'd died.

There were cases and cases of ammo in the tunnel, plus dozens of sleeping positions and boxloads of food. The enemy had prepared a real stronghold here, from where they could have withstood the longest of battles. At the far end of the tunnel there were side entrances, which had to connect to other tunnels and rat runs. But with the entire cave system having been pounded from the air, neither of us fancied taking a crawl further inside.

With the battle well and truly over, the OC ordered the Czech unit down from the high ground. The lads took their vehicles over to the south side of the canal. There they did a walk-around inspection of the woodstrip – the enemy position that the Apaches had malleted with cannon fire and flechette darts.

There they found an interlinked trench system with cleared arcs of fire, plus sleeping gear and caches of food and ammo. There were fourteen bodies in the woods, including one guy who appeared to be fixed to a tree trunk. He had tiny red stains all over his robe, and an AK-47 slung around his neck. He'd been peppered with flechettes, the tungsten darts nailing him to the tree.

Towards the western edge of the village one of the lads found the sniper point from which Sandy most likely had been shot. The enemy sniper had been firing through a tiny aperture. It was completely hidden from view, and we'd have had little hope of finding the gunman. There was one hole for the rifleman, and one for a spotter, just as our sniper units tended to operate.

A series of interlinked defensive positions were strung across Rahim Kalay. The OC had chosen this as the 'easy' route of advance into Adin Zai, as that's what the Intel had told him. In fact, this had been a fortress manned by hundreds of enemy fighters – complete with underground arms stores, bunkers, sniper holes, trench lines and a series of hidden tunnels to move around in, unseen from the air.

Had the Apaches not discovered that first compound position, and flushed the enemy out, we would have advanced on foot into the mother of all ambushes. Our attempt to take Rahim Kalay would have been met by a wall of death. It didn't bear thinking about how many of the lads of B Company would have been smashed in there. We would have lost far more than we had already that day.

As darkness crept into the silent village, we manoeuvred the Vector between two compounds that provided a little cover. We were to the north-east and on the far side of the village. To the north lay the open desert, and to the east stretched the Green Zone. Whatever enemy had survived the onslaught, it was into there that they would have fled.

It was 1900 by now, and for the first time since the day's battle had begun the company HQ element, the platoon commanders and the FST gathered together. In every soldier's mind was the same thought: that we'd lost Corporal Sandford, had two other serious casualties, and very nearly lost the battle. It was unbelievable that we hadn't lost more lads, and largely thanks to the Apache pilots that we hadn't.

Butsy talked for a while about Corporal Sandford's death, the way 6 Platoon were holding up, and what to do with Sandy's personal kit. Then he spoke of the assault plan, and what we might have done to prevent losing anyone. The OC was clearly gutted at losing one of his own. He took this very personally, almost as if it was his own failing.

'This is my analysis of what happened today,' the OC said. 'We stirred the hornets' nest. We stirred and we stirred and suddenly it erupted. What resulted was a long and intensive massive firefight. Air assets were made available as we needed them, and we couldn't advance until we'd neutralised pockets in that hornets' nest, and that's what we did.

'For an hour or more there was that massive, pitched battle,' the OC continued. 'Air, artillery and mortars were pounding the enemy positions, and back on the ground we still had troops in contact who needed water and ammo. We captured two-thirds of the village, and 4 Platoon were in the north trying to link up with the Czechs.

'That was today's battle.' The OC paused. 'Throughout all of this, the "what if" is could we have done anything to prevent Sandy's death? The only way to have avoided it would have been not to do the mission. And bear in mind the worst-case scenario: losing more soldiers during the initial firefight, which would have made it harder to get out.'

The OC took a thoughtful pause. 'Then, we would have come back in on foot and hit all of what the enemy had prepared for us here. It would have been a modern-day Rorke's Drift. We'd have been sucked in with no way out. We'd have been forced to fight our way out, taking massive casualties.

'We may have thought we were up against a rag-tag enemy,' he continued. 'We are not. Their ferocity and their ability to coordinate fire is clear. They have front-line trenches with cleared arcs; reserve trenches set fifty metres back, with sentry positions out front with RPGs. Their tactic has been to hit us, engage, fall back and draw us in to an ambush-and-surround position. We've been very, very lucky today.'

The OC gave orders that we were to consolidate our positions into all-round defence, and remain in Rahim Kalay that night. In the morning we would patrol out towards Adin Zai. And we would stay on the ground and dominate the territory.

After the briefing Major Butt did a walk around the company's positions. It took him two and a half hours, and he got to say a few words to every one of the men. He knew how that would boost their morale. He came back via our position, and Sticky doled him out a cup of coffee, which was the OC's chosen brew.

'Fantastic,' he remarked, as he took a sip. 'Just what I needed. Bloody fantastic, lads.'

I glanced at my watch. It was 0200 hours. I'd been curled up on the dirt next to the Vector, dozing, my TACSAT propped against my ear. I had no air, but I was scanning whilst I kipped, just in case I got anything unexpectedly.

We fully expected to get hit again that night, but I'd only get air if the enemy launched a full-on attack. In a way it was a good thing. I was totally exhausted, and all I wanted was to get my bracket down. As Butsy chatted away to the others, I drifted off into the sleep of the dead. Overnight, I kept getting woken by the odd crackle of gunfire, or the crunching explosion of grenades. I'd do a quick scan of the airwaves, but with no warplanes on station I'd drift off to sleep once more.

Dozing fitfully in between the worst attacks, I shook myself awake proper at 0400. I had air for the company stand-to. I got a pair of A-10s, and had them flying low-level shows of force over our positions. That way, if the enemy were planning anything major they could see that we had Warthogs on hand to smash them.

It was then that I learned just how fierce the fighting had been overnight. The OC described it as a night of 'enemy in the wire', with close-quarters fighting in the pitch darkness. Butsy hadn't slept a wink as he'd kept doing the rounds of his lads to bolster their morale. The enemy had probed us from every direction, and the lads had spent the dark hours wired, and wondering from where next the enemy were going to try to hit us.

One young soldier had told the OC about an enemy fighter who had simply refused to die. He'd shot the enemy fighter three times, but still the guy was trying to press home his attack. Eventually, the young lad had bayoneted him to death. It was clear the enemy were pumped up on drugs, for nothing else could have kept them going like that.

After stand-to I had the A-10s ripped by a pair of F-15s, call signs *Dude Zero Three* and *Dude Zero Four*. The F-15 was fast becoming one of my platforms of choice, particularly after the way that lone pilot had performed during the previous day's battle.

I got chatting to the pilots, and it turned out that one was a woman. Emma proved to be a friendly type, as the American girls often are.

'I haven't spoken to a pretty lass in weeks,' I told her. 'What you wearing up there?'

'I'm flying in my suspender set and bra, *Widow Seven Nine*. What you wearing down there?'

'I'm minging,' I told her, truthfully. 'I smell like a damp dog. And I haven't brushed my teeth for six days.'

She laughed. 'Gee – I guess that's what I can smell from up here then.'

The Intel coming down from the Rahim Kalay elders was that forty-one enemy had been killed, not to mention the wounded. Plus at least thirty-six enemy fighters were missing. We'd also taken out a full mortar team with mortar tube, and a further mortar team without the tube.

As if to confirm what the elders were telling us, the F-15 pilots spotted scores of tractors on the village outskirts, hauling out the dead and injured. We let them go about their grim work unmolested.

We would respect the enemy dead, and for sure they had enough of them right now.

THIRTEEN
THUNDER RUN

Two days later I was back at FOB Price, en route to Camp Bastion. For the next week at least my war was over. I'd been ordered to return to the UK, for – of all things – an AIDS test.

I was gutted to be leaving the 2 MERCIAN lads, especially as they were poised to take Adin Zai and occupy a swathe of the Green Zone. And I was pissed as hell to be leaving my FST, and handing over my JTAC role to a stand-in. But I had no option. Orders were orders.

Two months earlier I'd managed to stab myself in the leg with a discarded needle. The enemy often drugged themselves up before battle, shooting up with heroin or amphetamines. Their positions were littered with syringes, and of all the places I'd chosen to take cover I landed on a druggie's needle.

I'd been just six days in theatre, and the soldiers of 42 Commando had been ordered to take Sangin town on foot. They'd been on their way back to the UK at the end of their tour, when they were told to do one more mission. I got embedded with the commando, as the JTAC for the Sangin operation.

In briefings we were told this was going to be the biggest op of 42 Commando's entire tour. I'd never done a live drop before – controlling aircraft with live ordnance over a battlefield. I guessed this was going to be my baptism of fire.

Only three of us from the FST could go on the mission, and Sticky had drawn the short straw. We were embedded with Juliet Company, a bunch of kich-arse Commandos. There were 120

Marines in the company, and I was the JTAC in charge of the air. A lot of these lads were big, ugly, grizzled bastards, and I didn't think they'd take kindly to me dropping a bomb on their mates by accident. One of the commando's own JTACs came to have words with me. He was at the end of his Afghan deployment, and he was a qualified JTAC instructor.

'Bommer, when we go out on this op you're the JTAC heading it up,' he told me. 'Everything that's going to happen you're going to lead it. I'll listen in, and only intervene if I have to. You need to find your feet and find them fast. You OK with that?'

I swallowed hard. 'If there's anything, can I ask you the question, boss?'

He shook his grizzled head. 'No, mate. I'll step in if needed. I'm giving you your head, Bommer. You'll do all right, and if you don't I'll be on to you.'

So be it. I was in at the deep end.

We drove up to Sangin via the desert. The night before the assault we laagered up in the open, just short of the 611, the main road into town. We had C Squadron of The Light Dragoons – my parent regiment – in Scimitar light tanks doing overwatch of the road, to stop the enemy from planting mines or IEDs, or setting ambushes. The enemy knew of our intentions, and they were coming to join the party. Half a dozen top Taliban commanders had arrived from their northern stronghold of Musa Qaleh, each bringing sixty to eighty fighters with them. We were two companies of Royal Marines – some 240 men – up against several hundred enemy fighters.

The CO of 42 Commando gave us the final mission brief in our desert laager: 'Secure Sangin centre through shock action, moving from inside to out to secure. Gain positions with or without force; deny enemy firing points. Hold all until relieved by Task Force Fury.'

Task Force Fury was troops from the US 82nd Airborne's 4th Combat Team. They would be inserted in a massive heli-borne operation to the west of Sangin, with us coming in from the east in a pincer movement. At the same time units of The Light Dragoons would move in from the open desert, conducting a highly visible feint, in an effort to fox the enemy.

The CO finished his brief with these immortal words. 'Not all of you will be coming out of Sangin'. That drove it home: there were going to be a few lads getting whacked in there.

The airspace above Sangin had been formulated into a High Density Air Control Zone (HIDACZ), which was akin to an enormous ROZ broken down into individual sectors. We had a 'king JTAC' in control of the HIDACZ, and orchestrating the air from FOB Robinson. His call sign was *Widow Seven Zero*.

I got the alert via Chris that our forward unit had spotted an enemy mortar team setting up under cover of darkness. I got a description of the target, then radioed for air.

'*Widow Seven Zero, Widow Seven Nine*,' I put the call through to the king JTAC. 'Sitrep: visual mortar base plate setting up. Request immediate air support.'

'*Widow Seven Nine, Widow Seven Zero*, affirm. *Bone Two Three* is five minutes out of your ROZ.'

I had a B-1B semi-stealth bomber inbound. The American pilot came up on the air.

'*Widow Seven Nine*, this is *Bone Two Three*: request an AO update.'

'*Bone Two Three, Widow Seven Nine*. Sitrep: company-plus sized group stationary to the south of route 611, in overwatch of enemy mortar team setting up. Coordinates of mortar base plate are 59372057. Elevation 1,850 feet. Line of attack east-to-west. Nearest friendly forces four hundred metres south. Readback.'

The B-1B pilot read the coordinates back to me and confirmed the attack details. As we began the talk-on, I had to put myself into the mindset of the pilot in his cockpit. It felt just like being at JTAC school, only this was for real.

'This is what you're looking for,' I told him. 'There's a rectangular compound to the north of the 611. Just to the east of it is a small track leading north-east to south-west.'

'Visual with the compound and track,' the pilot confirmed.

'On the track directly to the west of the compound – one times white pickup parked under trees. Next to that, at nine o'clock: three times male pax, setting up mortar base plate.'

'Searching … Visual on the white pickup. Visual male pax. Preparing my attack run: what d'you want on target?'

'I want a GBU-38. Nearest friendlies four hundred metres south.'

'Affirm, one times GBU-38. Tipping in.' A pause. 'Sixty seconds to target. Call for clearance.'

As the giant bomber arrowed through the darkened sky, I heard a voice cutting in on my frequency.

'Break! Break! *Bone Two Three*, *Widow Four Six*: I'm now the call sign clearing you in to do this drop.'

What the fuck – I had another JTAC trying to break in and take control of my mission. Before I could say anything, the B-1B pilot was back on the air.

'*Widow Seven Nine*, this is *Bone Two Three* – who the hell's controlling this drop?'

'I'm the call sign with eyes on,' I told him. 'I'm the controlling station.'

'I'm over target without dropping,' the pilot announced. '*Widow Seven Nine*, that's an aborted drop. I'm coming around for a new attack run.'

'Roger that,' I replied. 'Stand by.'

I couldn't believe it. My first live drop, and it was a fuck-up due to some bastard JTAC trying to take it off me.

'*Widow Seven Nine*, *Bone Two Three*, banking around. Two minutes to my attacking run.'

Bringing around a B-1B was like turning the *Titanic*. It was a big beast of thing, hence the two minute delay. I'd told the forward unit of our company to pull back from the 611, to get a safe distance from the drop. They were no longer visual with the mortar team, so I had no eyes-on telling me whether it was a live target or not. It was all going to rat shit. The commando's JTAC got on the net to the guy who had been trying to take over my drop.

'Mate, you listen to me: you do that again and I'll get you sacked. You fucking leave the JTAC on the ground to do his job. You're not fucking jumping in. Got it?'

The JTAC on the other end said he understood, and that it wouldn't happen again.

Bone Two Three came up on the air. 'I've got eyes on the white pickup. Starting attack run now. Tipping in.'

As I waited for the pilot to call for clearance, I was double- and triple-checking the map, making sure I had my figures right, and that no other friendlies were in the vicinity of the blast.

'Clearance,' came the pilot's call.

'Nearest friendlies, four hundred metres south. Clear hot.'

'In hot.'

There was a pause for thirty seconds or so, as the monster bomber came in. I could hear the faint roar of jet engines at high altitude echoing through the skies. Had the mortar team also heard it, I wondered, and bugged out? There was no way of knowing.

'Stores.' The pilot gave the bombs-away call.

The GBU-38 was on its way. I was stood on top of one of the Viking armoured vehicles, with the commando's JTAC on my shoulder. We were watching for the flash of the explosion in the darkness. There was the howling scream of the munition cutting through the night skies, then nothing. It just stopped dead.

'Confirm you've released your weapon,' I radioed the B-1B.

'Affirm.'

'Well, nothing's happened. It's a dud. I'm requesting immediate re-attack.'

The pilot banked around, on another two-minute mega-turn. What were the chances of this happening? My first live drop, and a rival JTAC forced it to abort. Then the pilot drops a dud. The bastard mortar team would be halfway to China by now. The darkness and the terrain would have given them ample cover to sneak away.

I cleared the B-1B again. We heard the scream of the bomb coming in, then an almighty explosion lit up the night sky as it slammed into the desert just to the north of us. I had no idea if I'd hit that mortar team, but at least I hadn't smashed any of our own lads. It was my first live drop, and in spite of the fuck-ups it felt good. All the nerves and the fear were gone.

We did a final pre-assault check in the desert darkness, and I went through all the gear that I'd be carrying on my back. We had no idea how long we'd be in there on foot. I had to be fully mobile with all my personal gear, weapons, plus my JTAC kit. I had little doubt that I was carrying more weight than most of the commando lads.

Apart from my SA80 and my Browning, and all my mags, I carried the TACSAT, with the donkey dick aerial stuck of out the top of the pack. Then I had an infrared pointer device – like a maglite, but only detectable by night-vision – plus an LF28 Laser Target Designator (LTD), a bulky laser-firing gizmo. A few spare batteries for all the electronic kit really pushed the weight.

I gave a couple of extra batteries to Throp and Chris, for I knew the FST would remain together as a stick. With all the gear stuffed into my Bergen, I had forty kilos in there. I chucked in a few grenades – smoke, white phosphorus and high-explosive. I managed to squeeze in one ration pack, and three litres of water, plus some photos of the family, and that was it. I was chocker.

On my wrist I strapped a Garmin 101 GPS, a civvy device that I'd purchased in the UK. It'd cost me £109.99, and was probably the most useful bit of kit a JTAC could carry. Wherever I might be, it would give me a ten-digit grid reference of my position. It could display both latitude and longitude, and Military Grid Reference System (MGRS) – MGRS being the standard type of grid you'd pass to the air.

I had my name, blood group and ZAP number scribbled on my helmet in thick marker pen. I double-checked I had my St Christopher on the dog-tag chain around my neck, the one that Nicola had got for me prior to my coming to Afghanistan. St Christopher is the patron saint of travellers, and she'd made me promise to wear it at all times.

I was good to go.

At 0300 we set out under cover of darkness, a massive column of Viking armoured vehicles doing a classic 'thunder run' into Sangin in the pitch darkness. As we roared in I was thinking to myself: I've been on the ground a week; I've just completed my JTAC course and exams; I've been taught a trade and how to speak a brand-new language; and now I have 120 commandos relying on me.

As the convoy went in I had a pair of Apaches overhead malleting anything that looked even vaguely like it might be an IED with 30mm cannon fire. We reached Sangin centre and dismounted. We left a skeleton crew with the Vikings, and moved off on foot just before first light. We stuck close to the OC's HQ element, as the company pushed forwards. The town was eerily quiet. Apart from the dogs barking, there was an ominous silence. The dog howls went back and forth across the dark, deserted streets, and I just knew the threat was all around us. I could feel the enemy presence; sense it; touch it almost.

The lads hit the first compounds, and started going through them, bar-mining the walls, then chucking in frag grenades, followed by a burst of fire. The Apache above the forward line of troops, looking for enemy in the compounds, called in.

'*Widow Seven Nine*, *Ugly Five Three*: two male pax on rooftop position at governor's compound. They're in a tiny shed-like building, and they're watching your movements via binoculars.'

I reported it to Chris who passed it to the OC. He told us: 'Hit them.'

'*Ugly Five Three*, engage the OP position with one times Hellfire. Nearest friendlies three hundred metres west.'

'Stand by. Planning my run. Visual with two pax and preparing to fire.'

'Clear hot.'

'In hot.' A beat. 'Stores.'

The Apache came in over the top of us, the Hellfire blasting away from its launch rail. I saw it streaking in, and a second later it went straight through the small door of the shed and exploded inside, shredding it. All that was left was an angry ball of smoke billowing skywards, and a fringe of blown-up walling.

I'd called in my first strike of the mission and we'd killed some enemy. My adrenaline was pumping at eight million miles an hour. I was caught up in the action. I didn't ask the Apache for a BDA. I didn't need one. But he did have this for me.

'*Widow Seven Nine*, I'm visual with a large number of pax inside the main building below the target just hit. It's crammed full of them. Through the windows I can see weapons.'

'Roger. Stand by.'

The building beneath the destroyed shed was a large, concrete structure. Our lads would be advancing right past it, and it was an ideal ambush position. An armour-piercing Hellfire would go through it, not smashing it completely. I needed something bigger. I had a pair of Harriers stacked up above the Apache. I gave the pilot a call.

'*Recoil Four One, Widow Seven Nine*; we've got a build-up of pax in the compound at the target that Ugly's just hit.'

'*Widow Seven Nine, Recoil Four One*: I've been listening in to you and *Ugly Five Three*. Aware of the situation. Aware of the target.'

'What munition d'you recommend?'

'Due to the size of building and wall thickness – a one-thousand JDAM, with a ten-millisecond delay.'

It was a good call. A thousand-pound bomb with that delay would go through two floors before it detonated. I got the Apache to bank around north, so I could bring the Harrier right down on to target.

'*Recoil Four One*, friendlies are three metres to the west of target, behind hard cover. I want an east–west attack run, to keep the blast away from our troops.'

'Roger: an east–west attack run,' came the pilot's reply. 'I'll need two minutes to set up for my run. Stand by.'

I didn't have a great view of the target. I wanted to be dead certain if we were unleashing a thousand-pound JDAM. Our lads were danger-close. It was my third live drop, and if it had to be danger-close, I wanted eyes on target. I also wanted to lase the target, so the JDAM could home in on my laser beam.

'I'm going forwards,' I told Chris.

He nodded. 'With you.'

'Going forwards to get eyes-on!' I yelled to the OC.

'Right,' he yelled back. 'I'll hold the company stationary until the jets are done.'

I scuttled ahead, crouching down as much as I could under forty kilos of kit. Chris was right behind me, sticking close to my shoulder. There was a shallow alley that sloped away before us, rising up again to the target building. Hugging the walls for cover I pushed onwards, passing our line of forward troops. I was sweating like a pig and breathless. I could feel rivulets running down my back. I was also nervous as hell. I was in the middle of a warren of alleyways and mud-walled buildings. The clock was ticking, and I had to get this drop dead right.

I checked my watch. Fifty seconds to the Harrier starting his attack run. I crouched behind a compound wall that gave a little cover, and struggled out of my pack. I started chucking stuff out, as I scrabbled around for the Laser Target Designator. The LTD was about the size of a shoebox, and took up one hell of a lot of space.

Finally I had it. As I went to line up the target there was a burst of static on my TACSAT.

'Tipping in. Call for clearance.'

I grabbed the TACSAT. 'Roger.'

I went to lase the target, lining it up in the LTD's eight-times magnification sight. At the same time I was trying to double-check the map, making sure I hadn't missed any friendly positions, and keep an eye out for any enemy. The bloody LTD was getting in the way. It was just too bulky for this kind of fast-moving work.

We were a 150 metres short of the target, I could see the building clearly, and the pilot was only seconds out. I made the call to bring the drop in by visual means only.

'Clearance,' the pilot called.

'No change friendlies. I'm visual the target,' I confirmed. 'Clear hot.'

'In hot.' A beat. 'Stores.'

A couple of seconds later there was an ear-piercing scream, as the munition howled in. It came over our heads like a thunderbolt, a dark arrow shape streaking through the air at ninety degrees. It punched a hole clear through the roof of the target building, as if it were paper.

A split-second later came the massive detonation, the entire building erupting in an explosion of shattered concrete, flying bricks and dust. A vortex of smoke and debris blasted out in all directions, pluming a hundred metres into the air, and then a rain of stones and debris and shrapnel started crashing down all around us.

Chris and I locked eyes. 'Fucking hell.'

I got on the TACSAT. '*Recoil Four One, Widow Seven Nine.* Nice work. Send BDA.'

'Roger. BDA: anything that was in there alive is now dead.' There was a slight pause. 'Correction: I've got movement to the north-east of the compound.'

Before I could respond I got a call from the second Harrier. '*Widow Seven Nine, Recoil Four Two.* I can do immediate follow-up attack with CRV7 rockets.'

We checked with the OC that there was no change to the friendly positions. I cleared him in to attack. *Recoil Four Two* wasn't messing around. He unleashed eighteen CRV7 rockets, which saturated the entire compound in devastating explosions. Both Harriers did a follow-up BDA: there was nothing left moving now.

The battle for Sangin continued for that entire day, during which I did attacks using Apache, Harriers and A-10s. The enemy fought back with mortars, sniper fire, RPGs and small arms. By nightfall we'd set up position on the roof of an abandoned hotel in downtown Sangin, and most of the town was in our hands.

On the morning of the second day I was down in the hotel basement with Chris and Throp. It was like a bloody heroin refinery down there. There were big cauldrons full of brown gunk, and used needles everywhere. We'd heard stories about the enemy jacking themselves up on heroin, prior to battle; here was the evidence. One of the Marines had told us about an enemy fighter who'd taken a whole mag from an SA80 before he went down. He'd been high as a kite, and not registering the bullets as they tore into him. We were under sniper fire in the basement, and I'd just got into position to return fire with my SA80, when I felt a sharp prick to my leg.

I glanced down, and there was a used syringe sticking out of my thigh.

'Fuck me!' I yelled. 'I've got a fucking used needle stuck in me!'

Chris went and fetched the company medic. He took a look and told me there wasn't a lot he could do. I'd have to be tested for every kind of disease known to man, and – most importantly – for HIV-AIDS.

I was gutted, to put it mildly. I'd known when I deployed to Afghanistan there was a risk of getting shot or blown up. But I'd never even dreamed of catching HIV off a Taliban druggie's used needle.

That evening we got relieved by the 82nd Airborne. I handed over to their JTAC in the midst of a massive firefight. I had *Missip Two Five* and *Missip Two Six* – a pair of F-15s – doing an airstrike on an enemy mortar team with five-hundred-pound airbursts. I had them coming in 'shooter-shooter, swept right' – sixty seconds apart, both dropping ordnance, before banking off to their right.

At the same time Chris was calling in the 105mm field guns, which were pounding the enemy positions, plus we had rounds being lobbed in by the Marine's mortar unit. This was what an FST was designed to do – airstrikes, guns and mortars all at the same time – and the 82nd Airborne's JTAC had his eyes out on stalks.

I did a swift handover brief, then passed him control of the jets. 'Missip call-signs, *Widow Seven Nine* coming off station. Handing over to *Jedi One Six*.'

'Missip call signs, *Jedi One Six*: I'm now the ground controlling station.'

And that was our handover in contact. We did an eighteen-hour night drive back to FOB Robinson. En route Chris, Throp and I talked about what we'd say to Sticky. We felt sure we'd just been involved in the biggest battle of our tour (how wrong we were). The last thing we wanted to do was make anyone in the FST feel like an outcast. We agreed to play it down as much as we could.

I had a lot of time to think during that long drive. I had fifteen confirmed enemy kills, and I'd done a boatload of controls. I didn't feel like a veteran JTAC just yet, but I'd found my feet. Most importantly, I hadn't let any of the lads down. But it had all been spoiled by that bloody needle prick. It was preying on my mind. I was worried about that AIDS test. I couldn't even get checked right away: it would take months to develop in the body. The worst part of it was I'd have to go back to the UK to get tested.

Cuff – Corporal Grant 'Cuff' Cuthbertson, the JTAC who'd trained me – was out in the Afghan theatre. He'd listened in on the radio during my controls over Sangin, and he'd been all teared up at hearing me doing my thing.

'It was like teaching a kid how to ride a bike,' he told me, once I was back in FOB Price. 'Then seeing that kid go cycling off all on their own.'

I hadn't slept for four nights straight, and I couldn't wait to get my head down. But Sticky woke me around midday. Some American captain needed me to call him on a secure telephone. The guy's name was Captain Bouff Balm, or at least that what it sounded like. What kind of name was that for a soldier, I asked myself?

I was told the call was all about the Sangin op, and that it was urgent. Captain Balm was the US 'CJOC' based at Kandahar Airfield, whatever a CJOC might be. I called him on the secure line from the FOB Rob ops room.

'Sergeant Grahame, calling for Captain Bouff Balm,' I announced.

A voice came on the line. 'Sergeant, I just want to ask you some questions about the Sangin mission. You were the JTAC on that op, right?'

'Aye, I was.'

'Why did you start with a thousand-pounder and end up using CRV7? It's normally the other way round.'

'I used the thousand-pounder to get through the building's roof and flatten it,' I told him. 'The CRV7s were fired after and into the compound grounds, in response to movement around it. That's why I started big and went small.'

'Who was the nearest call sign to the compound when the strike went in?'

'I was.'

'Why did you say on the radio that nearest friendlies were three hundred metres away, when you clearly weren't? You were half that distance.'

'I was the most forward call sign. I had three walls between me and the target. If I'd made a mistake it would have been me who got flattened.'

'That answer's not good enough, Sergeant. If you lie to the jets on how close you are, you'll get someone killed.'

'But I was the most forward call sign, and I didn't really give a shit.'

'If it happens again I will take your qualification from you, Sergeant. And that comes all the way from a two-star general. Now, I want to talk to you about that thousand-pound drop.'

'What about it?'

'Well, it's come down from Intel that this is who you killed: Mullah Abdel Bari, Mullah Qawi, Mullah Hafiz and his bodyguards. That's three top enemy commanders. So fucking well done. That comes from the general – it was an excellent strike.'

I stared at the receiver for a second in sheer amazement. 'Hold on a minute, mate: if it was that good a drop why are you going on about taking my qualification away?'

'You're new on your tour. We've got to bed it into you how to operate. Don't say you're a distance away when you're not. And don't mess around with gettin' danger-close to your own drops.'

'Fair enough,' I told him.

Captain Bouff Balm seemed to think 150 metres was too close for comfort. By the end of my tour I'd be dropping bombs at a fraction of that distance, on a daily basis. It was the only way to keep myself and the lads alive.

FOURTEEN
OPERATION
MINE STRIKE

The AIDS test in Norwich Hospital was a negative, which was a massive relief. Whilst awaiting the result I'd gone to see Nicola and the kids. I tried telling her as little as I could about what I'd been up to in theatre, but she's a canny lass and she knew I'd been busy out there.

'So, have you killed anyone?' she asked me, eventually.

'A couple,' I lied. In truth I'd lost count.

I then proceeded to do exactly what the kids wanted every minute that I was home. But the hours passed in a blur, and it was a total headfuck being back in the UK and trying to act as if everything was normal.

Everything wasn't normal. The lads were out in the Green Zone embroiled in the fight of their lives. And I couldn't help but think that's where I should be, bringing in bombs on the enemy to smash them. Before leaving I tried explaining to Nicola the importance of what I was doing out in Helmand.

'I've done all right out there, trying to look after the lads,' I told her. 'The air's a massive part of the picture. And that's what matters – looking after the lads. Not killing people.'

'I know. I know.' Nicola was all teared up. 'Just don't take any stupid risks, OK? I know what you're like, Paul.'

That was the great thing about Nicola: she never said 'don't go'. She knew what I did for a living, and she never once tried to stop me.

Every time I went to war we both knew it might be the last, but she never once tried to stop me. What a top girl.

I made her a promise. 'When I'm home, we'll go to Disney for Ella's second birthday. Soon as I'm home. I reckon we'll have earned it.'

Three days later I was back in Camp Bastion and ready to rejoin the lads. I was also itching to catch up on all the news. Major Butt had taken B Company through the enemy positions, driving them out of the Green Zone. Rahim Kalay and Adin Zai were now owned by our boys, but it had taken days of intense close-quarter fighting to secure them. One of those actions had already passed into 2 MERCIAN legend.

The OC had sent a fighting patrol into the Green Zone at night, to winkle out the enemy. At around 0200 they had been challenged in Arabic. The enemy opened up with a barrage of RPGs and machine-gun fire. Using a drainage ditch full of human faeces as cover, the platoon withdrew from the firefight. But in the confusion of doing so one of their number, an eighteen-year-old private, had gone missing.

The patrol was led by Sergeant 'Jacko' Jackson. Jacko had called the young lad repeatedly on his radio. Finally he'd answered. He was lying in a ditch with his leg shot up. He'd put tourniquets around it, and was clutching a grenade in either hand. As the enemy got ever closer he'd pulled the pins out with his teeth, and was holding the release levers closed. He was going to blow himself and them sky high if they found him.

The patrol couldn't see where he was, so they asked him to throw an IR Cyalume – an infrared light stick visible only by night-vision – into the air. That would give them a fix on his position. The trouble was the private had a grenade in either hand with the pins out, so how was he going to throw the light stick? He opted to jam one of the grenades between his knees, and use the free hand to hurl the Cyalume.

It was at this stage that the private said he was visual with the platoon ten metres away. He presumed the lads had come to rescue him. In fact, the platoon hadn't moved out of their shit-filled ditch, so it had to be the enemy. Sergeant Jackson ordered the wounded soldier to cease talking on his radio, for it was leading the enemy to him. The private replied that he was going to blow the grenades, if that was the enemy so close to him. Jacko managed to talk him out of it, and promised that they were coming in to get him. He then decided that they needed a feint, both to distract the enemy from the Cyalume throw and the rescue attempt, and to draw their fire.

Sergeant Jackson opted to lead a lone assault on the flank of the enemy position, so they would think they were being surrounded. He set off into the darkness, crossed the terrain and charged the enemy with assault rifle blazing, and chucking in some grenades for good measure. As Jacko had gone mental with his weapons, the rest of the platoon saw the Cyalume go up and rushed in to the rescue. They got their wounded man out, and Jacko managed to extract without being killed or captured. A week later the lad was back in hospital in the UK, with his leg patched up and trying to chase the nurses.

Jacko had tumbled over at one stage during the attack, as something powerful smashed him to the ground. The following day the OC had been giving him a bit of a bollocking, 'cause he couldn't raise him on the radio. Jacko had shown the OC his radio kit, and pointed out that the green light was on, proving that it was working. In response Butsy had pointed to a bullet hole that went straight through Jacko's radio, his backpack and into his body armour. No one had noticed it up until then.

'Look, Jacko, that's why your bloody radio's not working,' he'd remarked.

But in taking Rahim Kalay and Adin Zai we'd sadly lost another of the lads. Guardsman Daryl Hickey had been part of Somme Platoon – the same bunch that had got blown up in the 107mm rocket strike on their Vector. The Somme lads had been providing covering fire, as B Company assaulted an enemy position. Guards-man Hickey had taken a gunshot wound. He was evacuated by Chinook, but was dead on arrival at Camp Bastion.

I couldn't help but wonder if I could have saved him, had I been the JTAC on the ground during those battles. It was an irrational thought, and there was no reason to think my replacement was any less capable than me. But I couldn't help thinking it, and it only served to heighten my hunger to get back on the ground with my team.

I was told that I was leaving the following morning for FOB Price, but I was going by a slight detour. I was to run an Operation Loam resupply convoy to Sangin, some fifty kilometres north-east of Camp Bastion. Every month an Op Loam convoy would set out, a line of trucks with escorts seeking to deliver food, ammo and water to a string of British bases along the Helmand River valley.

The convoys ploughed through the open desert, throwing up a massive plume of dust. They were visible for miles around, and they were forever being targeted by the enemy. Those convoys had earned the less-than-affectionate nickname of 'Operation Mine Strike'. This month's Op Loam convoy was short of a JTAC, so muggins here got the tasking. En route back from Sangin I'd get dropped at FOB Price, and link up with my FST.

I did not want to do this, but at least it'd get me back with the lads. I had no gear whatsoever in Bastion, so I had to beg, borrow and steal a JTAC-ing kit. I went to the Signals Compound, and got myself a TACSAT. They filled it up with the crypto for me – the encrypted signals information I needed for it to work, and I blagged myself a rifle and a pistol.

At 0400 the convoy left Camp Bastion. In charge of the convoy was the packet commander – I was riding with him in his Vector. Behind us were fifty-two MAN fifteen-tonne army trucks, with WMIKs, Mastiffs and Vectors in support. It struck me that these logistics boys – the 'loggies' – were the unsung heroes of this war.

There was no glory in running this gauntlet every month, and it was certainly no fun for anyone but the enemy.

Seven hours after setting out we had our first vehicle go over a mine. A WMIK had been roaming out front, clearing the way ahead. The massive explosion blew a crater the size of a house in the sand, and blasted the WMIK thirty metres or more from where it had been hit. The WMIK Land Rovers have mine protection provided by ballistic matting, and the two guys in the front were pretty much OK. The rear had taken the brunt of the blast. The 50-cal gunner had taken shrapnel in the leg, and his left foot was in a bad way. The convoy medic put a tourniquet around the guy's shin, but there were thick globules of blood oozing out of the seams of his boot. We had to get the guy back to Camp Bastion, so they could cut the boot off him, and treat the injury. I got on my borrowed TACSAT.

'*Widow TOC, Widow Seven Nine*. Sitrep: I'm with the Op Loam resupply convoy thirty kilometres due west of Lashkar Gah. We've got a WMIK hit by a mine, and one T2 casualty. I need immediate air, plus IRT. I'm making up my own ROZ: ROZ Bommer.'

I needed a ROZ, so I could be free to control air missions above the convoy. I got allocated a pair of F-15s inbound, *Dude Zero One* and *Dude Zero Two*, plus a Chinook with Apache escort to do the casevac. ROZ Bommer was getting busy.

We got the casualty on to the bonnet of a WMIK, so we could drive it up the ramp of the Chinook and into the aircraft's hold. The medic had the lad's shins strapped together, to hold the injured leg still, and he'd got some morphine into him to stop the screaming.

First into the overhead were the F-15s, and I got them flying air recces above the convoy. It wasn't two minutes before they spotted trouble.

'*Widow Seven Nine, Dude Zero One*: I have a compound five hundred metres to the south of your position. I'm visual with forty males of fighting age. No weapons visible, but it looks like an ambush.'

'Roger. Keep a close eye on 'em. Wait out.'

My first priority was to get the wounded on to the Chinook, and the convoy moving. Sat in the open desert like this we made a peachy target. I also had to deny the WMIK to the enemy, not that there was much left of it. I'd get that done whilst awaiting the Chinook.

The wrecked vehicle was way to the front of the convoy. I asked *Dude Zero Two* to hit it, whilst his wing kept an eye on the compound. The packet commander was a young lieutenant, and he was giving a running commentary on what was happening to his OC, back in Camp Bastion.

I'd just begun my talk-on with the F-15, when he interrupted me.

'The major says do not at all costs deny the WMIK. We need to recover it to Camp Bastion.'

I stared at the guy for a second. 'Tell your major to fuck clean off. We're not recovering it. We're blowing it up.'

The guy relayed my message. '*Widow Seven Nine* says negative – he's going to blow it up.'

I heard the major kicking off on the net. 'Well, tell *Widow Seven Nine* if he does, I'll be billing him the cost of a replacement vehicle.'

'Listen, mate: you brief your boss the following,' I rasped. 'We've got forty males of fighting age gathering in a compound ahead of us. The Dude call sign says it's an ambush. The WMIK is totally fucked. We need to get the casualty out, and the convoy moving.'

The lieutenant relayed my message, and the major repeated his order to recover the WMIK. He wanted it lifted on to a truck and taken back to Bastion, so he could check for himself if it was beyond repair.

I grabbed the lieutenant's radio. 'This is *Widow Seven Nine*: I'm no mechanic, but Jim couldn't fix the bloody thing. It's fucked. I need to blow it up and get us moving.'

'I repeat what I've said: I want that vehicle recovered.'

'You want me to jack you up a helicopter, so you can fly out yourself and see how fucked it is?'

There was no response. It sounded like the major had switched off his radio.

So be it. I'd been here before on this tour – when the Vector had been hit by the 107mm rocket, at Adin Zai. Plus we'd had WMIKs hit in mine strikes. In each case any attempt to recover kit or weaponry or the vehicles themselves had been met by savage follow-up attacks. We'd learned the hard way that putting lads at risk for a knackered vehicle wasn't a clever idea. But somehow, I doubted whether the young lieutenant in charge of the convoy had such combat experience.

I turned to him. 'Right, I'm blowing it up. You can blame me if there's any comeback.'

The lieutenant looked doubtful. 'We can't do that. There'll be hell to pay.'

'Listen, you can just blame me. We've got to get this convoy moving.'

He shook his head. 'I can't do it. I'm going to get the recovery truck forward to begin lifting it.'

I gave the guy a look, then dialled up the IRT. It was his convoy, but it was a dumb decision if ever there was one. More importantly, he was putting his men's lives needlessly in danger.

The IRT Chinook was three minutes out. I got one of the lads to mark the LZ. He dropped a green smoke grenade on the patch of open ground. The surface of the desert was covered in a fine, talcum powder dust, and as the Chinook came down we were engulfed in the massive brown-out of a rotor-driven dust storm. We got the WMIK up the ramp, the wounded guy onboard, and the heavy in the air again in two minutes flat. That done, I tried again to persuade the lieutenant to let me blow the WMIK. He refused.

Whilst the recovery truck came forward to lift it, I got the Dude call signs flying shows of force over the compound to our front, firing flares. I had no doubt those were the bastards who had planted the mine, and they were there just itching to launch a follow-up attack. The F-15s flew repeated passes twenty metres above the compound, as the low-loader manoeuvred into place. We had no mine-clearing kit, and without doubt the enemy would have planted more than just the one – so every man engaged in the recovery was risking his life. It took forty-five nail-biting minutes to recover the wreck, every second of which I was expecting to see a vehicle or a bloke go over a mine and get blown to fuck. Thankfully, luck was with us, and we got on our way with no further casualties.

As we pushed onwards we were only making fifteen to twenty kilometres an hour. Trucks kept getting bogged down the whole time, whereupon they had to get towed out of the shit by another wagon. It was a hideous way to travel, and I was mightily pleased to hit FOB Robinson, at 2200.

The packet commander planned to do the run up to Sangin in the early hours, to catch the enemy napping. In the meantime, we had the chance to get some kip.

I awoke at first light to discover that the convoy had left without me and the packet commander. Somehow, they'd managed to get on the road for the run down to Sangin without their commander or their JTAC.

The convoy made it back to FOB Robinson without being ambushed. The plan was to set off at 0600 on a non-stop drive to Camp Bastion. But we'd had Intel come in that the enemy were planning to smash us big time on the return journey. We knew roughly where the attack was expected, and I helped the packet commander cobble together a new strategy.

I got in touch with a fellow JTAC, Sergeant 'Bes' Berry, one of the guys who'd trained me back in the UK. He was the JTAC embedded with a unit of Estonian Commandos, who were operating in the high ground to the east. We wanted them in overwatch of our route around the expected ambush point.

Bes said he was fine with that, but that he had a slight problem. Whilst tabbing up the mountain one of the Estonian lads had ripped the sole off his boot. He'd cut a lump out of his roll mat, and black nastied (gaffer-taped) it to his boot. If we could get the guy a replacement pair of size sevens, they'd happily do the overwatch of our convoy. I got on the radio, and got a replacement pair of Meindls put on that morning's Chinook flight down to FOB Rob. I radioed Bes with the news.

'Mate, the plan is we'll rendezvous with you en route. I'll pass you the boots, if you keep an eye on us lot, OK?'

'Good one, Bommer, you got a deal,' Bes replied. 'By the way, mate, how was the AIDS test?'

Everyone seemed to have heard the story about my 'little prick', as they were calling it.

'I'm riddled, mate. Only joking. See you tomorrow for the boot drop-off.'

The Chinook arrived with a pair of Meindls, but they were size nine. I spoke to Bes, and asked him what he wanted: a brand-new pair of size nines, or the sevens off my own feet. They were in fairly decent fettle, if a bit smelly. He said he'd take mine.

At 0400 the convoy set off. I had *Vader One Seven* and *Vader One Eight* in the overhead, a pair of Lynx helicopters. The Lynx has problems operating in the intense heat of the Afghan day, so I rarely got it as a platform. But it was early and still cool enough for the Lynx to fly. I got the Lynx flying air recces ten kilometres to the front of the convoy.

At 0530 I got the call.

'*Widow Seven Nine, Vader One Seven.* I'm visual with two male pax on a motorbike, ten clicks ahead of you. Every fifteen metres or so they're stopping, removing their backpack, and laying things on the ground.'

'What d'you reckon they're up to?' I asked.

'It's clear as anything: they're mining your route.'

'Can you see any weapons?' I asked.

'Negative.'

'Roger. Wait out.'

I radioed Bes. From his mountaintop position he was visual with us, plus the enemy motorbike team. Bes confirmed what the pilot had already told me: the motorbike boys were laying mines.

I put a call through to Widow TOC. 'I have Vader call signs ten kilometres ahead of my convoy, visual with two male pax on a motorbike. Every few metres they're stopping and laying mines on our route.'

Widow TOC replied: 'Unless you can see weapons under no circumstances are you to engage. If they're unarmed, you cannot engage.'

'Well, it's hardly Cleveland bloody County Council fixing the motorway, is it?' I replied. 'What's mines if not arms?'

'Repeat: if you cannot PID weapons you cannot engage.'

I came off the radio fuming. Thirty minutes later I lost the Lynx, for the air temperature was getting too hot for them to fly. Ten minutes after that we made the rendezvous with Bes's unit, and I handed over my boots.

'Your lad's welcome to 'em,' I told Bes. 'No way is he getting off that mountain with a lump of carry-mat wrapped around his foot. And thanks for keeping overwatch, mate.'

'No worries,' said Bes. 'I'll buy you a pint in return. Stay safe, mate.'

We pushed ahead. Some four hundred metres beyond where the Lynx had spotted the motorbiking pair, we hit the first mine. A WMIK out front took a blast bang under the front wheels. The vehicle wasn't as badly damaged as the first WMIK, but it was well out of action.

Worse still, all the crew were injured. There were broken arms and lacerations, and one of the guys had damaged his back. I was bloody seething. We'd had the minelayers in our sights, and the Lynx could've nailed them. But I'd been ordered not to fire. As a result we now had three lads seriously injured, and a fighting vehicle half torn to pieces.

What was the sense in any of that?

I got the IRT called out, and I was allocated a pair of Harriers. I then got a call from *Ugly Five Zero* and *Ugly Five One*, the Apache pilots that had done such a fine job of smashing the enemy over Rahim Kalay. They'd been monitoring the air. They were en route back from a mission, and could offer me thirty minutes' playtime.

I got the Harriers ramped up high, and the Apaches in low searching for ambush teams. I was itching to find the bastard enemy and smash them. I got the Chinook in to do the casevac, and the three wounded lads were loaded aboard. There was no sign of any hostile presence, so the convoy pushed ahead with Apaches and Harriers in overwatch.

Almost immediately, a beast of a Mastiff armoured vehicle hit a mine. It had a shredded tyre, but that was about the only damage. I called the same Chinook back in again, and this time it lifted off with one of the lads from the Mastiff. The injured lad had been knocked off his perch by the shockwave of the explosion, and broken some ribs.

It was 1900 hours by the time we reached FOB Price. This was where I was getting off. I told the packet commander I'd listen in on the radio during their drive to Camp Bastion. It was a well secure patch of road, the air would remain over the convoy, and I'd direct any controls if need be. An hour later the convoy was safely back at base.

It was a joy to be reunited with my FST. Well, kind of. Chris, Sticky and Throp gave me a good ribbing about my 'little prick'. Once we'd got that out of the way, we got down to business. We were leaving FOB Price first thing the following morning and heading out to the Green Zone.

We were making for three new bases at Rahim Kalay and Adin Zai. Patrol Base Sandford – named after Sandy – had been established on the ridge line overlooking the jungle. To the east lay

Monkey One Echo, a second fortified compound set at the limit of our eastward push into enemy terrain. But the real jewel in the crown was Alpha Xray, set right in the heart of the Green Zone, on the verges of the Helmand River.

Together, the three bases formed 'the Triangle', an area in which we were to find, fix and kill the enemy. The Triangle served as a chokepoint, restricting the flow of men and weapons throughout the length and breadth of the Green Zone. As such, it was the front line between our big garrisons at Gereshk and all stations south, and the enemy.

We were going in to hold that front line for anything from two weeks to the entire remainder of our tour. *Get in*. I had one crucial thing to do before packing all my gear and getting some kip. I had to ring my dad and wish him a happy birthday.

As I dialled the number, I reflected on how the two of us had parted on not the best of terms. I hoped he was over it, for I didn't want any rifts in the family. A few days before leaving for Afghanistan, I'd called to have a chat with him. We were close, and I didn't do much in life without talking to him about it first. At first my dad had never believed I'd make it in the Army, but that was well behind us now. He understood that the Army was my life, and he was proud of what I was doing. In passing, I mentioned I'd just taken out some extra life insurance for the wife.

'Why're you doing that?' he asked. 'The Army provides life insurance, don't they?'

'Yeah, I was just getting some extra. A top-up, like.'

'But why're you doing that?' he asked again.

'Well, just to make sure she's OK. She'd have the mortgage paid off and a big lump sum, so she'd be all right in life.'

'But why're you doing that?' my dad asked for a third time.

'What d'you mean – why am I doing that?'

'You're just wasting your money, aren't you?'

'No. Not if owt happens.'

—

'What's going to happen?'

'Dad, I'm going to Afghanistan.'

'What d'you mean by that?' he asked.

'I mean it's Afghanistan, and bad things happen in Afghanistan.'

My dad was worried, but I told him and my mum that soldiering was my job, and that I had to go. I hoped, with a happy birthday phone call from Helmand, I could put all of that to rights. Sure enough, Dad was over the moon to hear from me. The life insurance thing wasn't mentioned. I guess he'd reconciled himself to the fact that his son had chosen to go into harm's way to get a job done.

There was no way we could let a bunch of medieval lunatics like the Taliban take over an entire country, less still to use it as a base for global terrorism. That was what we were fighting to prevent. The hard-line Taliban hate our way of life. They want the world to be like Afghanistan under their rule. The Afghan people deserve more. We all do. I'd fight to the last man to prevent my family and my fellow countrymen ending up in a world like that. My dad hated me risking my life, but at the same time he understood what I was doing.

Sticky, Throp, Chris and I joined a resupply convoy heading out to the Triangle. We arrived at PB Sandford at night, after a hard day's driving through hostile terrain. We'd made painfully slow progress, as we checked for IEDs and ambushes all the way. Upon reaching the darkened base, Butsy told us we'd have our first briefing after stand-to.

And I felt like I was home again.

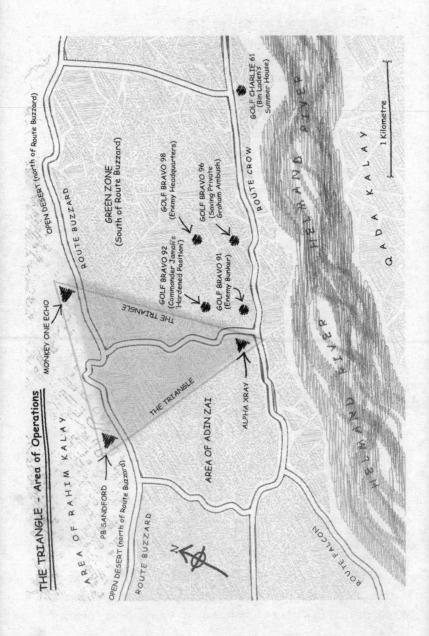

THE TRIANGLE – Area of Operations

AREA OF RAHIM KALAY

OPEN DESERT (north of Route Buzzard)

ROUTE BUZZARD

MONKEY ONE ECHO

GREEN ZONE
(South of Route Buzzard)

GOLF BRAVO 98
(Enemy Headquarters)

GOLF BRAVO 96
(Saving Private
Graham Ambush)

GOLF BRAVO 92
(Commander Jamali's
Hardened Position)

GOLF BRAVO 91
(Enemy Bunker)

GOLF CHARLIE 61
(Bin Laden's
Summer House)

ROUTE CROW

THE TRIANGLE

THE TRIANGLE

AREA OF ADIN ZAI

ALPHA XRAY

PB SANDFORD

OPEN DESERT (north of Route Buzzard)

ROUTE BUZZARD

ROUTE FALCON

N

HELMAND RIVER

QADA KALAY

1 Kilometre

FIFTEEN
THE KILLING BOX

The only change about Major Butt since I'd last seen him was his weeks-old stubble, and he was noticeably thinner. The well at PB Sandford was infested with bloodworms, and bottled water was too precious to waste shaving. By default, battle dress in the Triangle was going to be big beards all around. In his usual, gruff manner, the OC got down to business.

Butsy indicated a section of terrain on the map that he'd spread before us. 'Lads: we kept mowing the grass, and it kept growing back again. We'd push the enemy out of here, and they'd creep right back in. So, now we've pushed the enemy out once and for all. The feature we can easily recognise as our front line is the Adin Zai road. That's here.'

The major traced a thin line on the map snaking through the Green Zone.

'The operational concept behind our bases is as follows.' The OC glanced in my direction. 'Since you were last here, we pushed out from Rahim Kalay and took the Triangle. We pushed out of Rahim Kalay, using a double hook: B Company in the north, and soldiers from the Afghan National Army in the south. We hit massive resistance, and it was there that we lost Guardsman Hickey.'

The OC paused. 'The ANA lads did well, and were supposed to remain with us. Unfortunately, they were pulled away. So whereas it was supposed to be two companies holding the Triangle, we are now one. I've had to work out how best to use B Company to dominate the area. Our brief is to hold the terrain, and block the enemy out of Rahim Kalay and Adin Zai, enabling reconstruction and security in the areas we've secured.

'We have four platoons: one at Alpha Xray, one at Monkey One Echo and a platoon-plus here at PB Sanford. The platoons are on four-day rotations around the bases. In any of the three we are effectively surrounded. We've been taking hits just about every night, and at times we've been under threat of being overrun – especially down at Alpha Xray. I'm going to start using good old fighting patrols at night, to keep the enemy on their toes. We have to take the initiative, and the battle, to the enemy.

'FST ops are unchanged,' the OC continued. 'You're to use the air, guns and mortars in support of the lads, to smash the enemy. We faced fierce resistance getting in here, and we've been very busy since. You will have your hands full. It's the same-old same-old.' He glanced up and gave us his tough smile. 'And Bommer – you're to do whatever you want with the air to look after the lads here in the Triangle.'

'Yes, sir.' I was back in action as B Company's JTAC.

I was on a steep learning curve. Before leaving FOB Price Damo Martin had got me into the Fire Planning Cell (FPC) tent and handed me a bundle of new maps. The GeoCell had done a sterling job of compiling mapping that could be passed to the pilots on disk, and uploaded on to their weapons computers. The same maps were provided to us JTACs.

Most importantly, those maps had just about every known enemy position marked on them, with a codename. Enemy positions in the Triangle had been given the Golf Bravo prefix. So, for example, a tree-line might be Golf Bravo Nine Three. In theory, all I needed to do to talk the air on to target was give the pilot the GeoCell codename. It was a great concept: I wondered how it would it work in practice.

As I glanced over the OC's maps, the strategy behind the three bases in the Triangle began to make sense. Alpha Xray was a simple, mud-walled compound reinforced by rolls of razor wire, and with sandbagged rooftop positions. It was located in the heart of Adin Zai – in what had recently been prime enemy territory. It was the bait to lure them out to fight us.

PB Sandford was set a kilometre back from Alpha Xray, on the ridge line. It was a large compound, with thick mud-brick walls reinforced by HESCO barriers. From the base, the platoons had eyes on the Green Zone rolling out below, including Alpha Xray (AX). AX was linked to PB Sandford by two small dirt tracks that snaked their way through the bush, enabling convoys to fight their way through and resupply the base.

Monkey One Echo was 800 metres due east of PB Sandford, and set on the lip of the high ground. It was thrust against the limit of known enemy territory, and was our second provocation. The thick bush of the Green Zone ran right up to the walls, so it was almost as vulnerable as Alpha Xray. From Monkey One Echo a huge swathe of enemy terrain was visible, and the base was linked to PB Sandford by a dirt track, enabling resupply. The three bases formed a kill-box, in which to trap and smash the enemy.

After Butsy's brief we moved down to Monkey One Echo (MOE), our home for the coming days. MOE was another, massive compound with thick mud-walls. It was owned by a wizened Afghan elder who refused to be cowed by the Taliban.

Trouble was, we weren't to go inside the compound. We knew if we did the Taliban would really start to target the old boy's family. Instead, the lads had made a camp against the outside back wall. That wall was high enough to provide cover for our Vector. With a camo-netting lean-to slung against it, at least there was some shade. The more fortunate lads had US Army camp beds – a fold-out aluminium frame with canvas covering. Those like me without dossed down on the dirt.

From the vantage point on top of the compound wall the view over the old boy's compound was spectacular. In the foreground there were the dome-roofed buildings of his family home, complete with a sculpted mud arch like a half-doughnut. That arch led into a massive, three-storey square tower, with spyholes. It looked like something out of *The Arabian Nights*, and it had to be centuries old. If walls could talk, that place would have some stories.

To the west of the compound lay PB Sandford, and to the south-west Alpha Xray. Everything to the south and east was bandit country. To one side of the compound was a concealed observation position (OP) – little more than a crater scraped in the hard dirt, with some camo-netting slung over it. It was large enough for the FST to crawl into, and keep eyes-on below.

Once we'd set up camp, I had a chat with the JTAC who'd stood in for me whilst I'd been away, a guy called Stu. He briefed me on what had been going on in the Triangle. I was itching to get the handover done with, and get back in the hot seat with 'my lads'.

Stu talked me around a new bit of kit that he'd been using – a Rover terminal. It was the size of a small laptop, and it enabled the JTAC to see what the pilot was seeing, in real time. It fed off a live downlink from the pilot's sniper optics, which beamed down the video images. He passed me the Rover terminal as part of the handover.

'It's a top piece of kit, mate,' Stu reassured me.

'Oh fuck aye,' I told him. 'If it does what you say, it's mint.'

Stu talked me through my ASRs – I had air booked at first light and last light for the next three days. Those were proving to be the enemy's favourite times to hit us. And that was pretty much it: handover done. Stu got a lift back to FOB Price with the resupply convoy that we'd come in on, and we were back in action.

I got my first chance to use the Rover terminal that evening, when I got a lone F-15, *Dude One Five*. I had him for an hour, and it left me in no doubt as to the value of that Rover screen. I gazed in to the blue-green glow, seeing everything the pilot saw as he viewed it, in up-close detail. As he talked me through the terrain, I was right there with him. It was an awesome bit of kit, and I couldn't wait to see how it performed in a combat situation.

There was a well nearby, and that afternoon I'd washed all my kit. After doing the Op Loam convoy, and the journey in here, everything was minging. The well water came from a good twenty metres down, and for the split second that I chucked it over myself I was cool, in spite of the 46°C heat. I put my combats back on soaking wet, in an effort to keep the heat down.

Butsy was camping out with us at MOE, and he came over to have a quiet word at the well. The OC told me he was chuffed as nuts to have us back again. We weren't from his regiment, but we'd become like his boys. It was good to have the A-Team back together again. We were B Company's FST, and I was their JTAC.

That night I dossed down on my blow-up bedroll on the dirt floor. It was mid-summer, and there was no chance of rain. No one bothered with mozzie nets, although there was something that kept biting. I drifted off to sleep with the Afghan stars for company, and feeling pretty damn happy to be back where I belonged.

At 0430 the enemy hit us. Everyone was up at stand-to and ready to rumble. From out of nowhere, our position was raked with small arms fire. I heard the crack that high-velocity bullets make – a sound that had become so familiar over the last few weeks – as rounds went whining over the wall of the compound and slamming into the open desert beyond. At the same time I could hear PB Sandford getting smashed, so it was a two-pronged assault. I had an F-15, *Dude One Three*, overhead, but there was nothing to be seen. Via the Rover terminal downlink, I could appreciate the problem the pilot was facing. Wherever the enemy were hitting us from, they were well hidden in the dense bush of the Green Zone.

Once we'd beaten off the enemy attack, Sticky and I headed out to get a feel for the lie of the land. The compounds to either side of us were deserted, but in one we discovered these bulging hessian sacks. Inside each was a sticky, dark sugary substance – so-called 'brown' – the first step in refining opium poppy into heroin.

On the windowsill above the sacks was a string of Muslim prayer beads. It struck me as pretty rich the way the Talibs combined their hard-line version of Islam, and heroin. Somehow it was wrong to drink beer, but OK to bankroll a war via hard drugs. The contents of those sacks had to be worth a mint. We took the sacks, sliced them open with our bayonets, and emptied the contents into the river that ran past our position. I felt good doing it. That way, no one would be using that brown shit to buy bullets or bombs to mess up any of our lads.

A *shura* was in the offing. The OC wanted to explain to the local elders – blokes like the one whose compound we were camping out at – what we were doing here. As per usual we had Intel that the enemy would attack, just as soon as the *shura* was over.

I had a pair of A-10s and a pair of F-15s overhead. I cleared it with the OC that I'd get them to fly a low-level show of force over the meeting, then bank them up high so those at the *shura* were free to talk. I got all four aircraft in low and fast with flares, then banked them off flying air recces all around. It was an awesome show of force, and it must have put the wind up the enemy. Our team got in and out unmolested, but we had picked up some interesting enemy chatter. Commanders were ordering their men to stay in their 'hidden positions, to avoid getting seen by the aircraft'. They were telling their men to await orders to attack and kill the 'infidels'. I guess that was us, then.

The first, probing attack came at the enemy's favoured time – first light. Overnight, a raging sandstorm had blown up to the south of us. I didn't have any fast jets, for nothing was able to get airborne.

All I had was a Predator Unmanned Aerial Vehicle (UAV) orbiting above. The MQ-1 Predator does carry one Hellfire missile under each of its glider-like wings, but it is basically designed for surveillance tasks. At eight metres long, it orbits at 20,000 feet, the propeller to the rear of the aircraft driving it along at a cruising speed of around 130 kilometres per hour. The UAV is invisible from the ground, and totally silent. As a stealthy spy-in-the-sky it was an awesome piece of kit, and it had a fantastic downlink to my Rover terminal. But a ground attack aircraft it was not.

At 0530 all hell broke loose down at Alpha Xray. First came the angry crackle of small arms fire, as the enemy opened up on our version of the Alamo. Then came the swoosh-boom of RPG rounds slamming into the base. An instant later the platoon were returning fire, with their 50-cal heavy machine guns chewing into the enemy positions. The thump-thump-thump of the big guns was interspersed with the noise of SA80s firing off on single shot, and the boom of grenades from the lads' underslung launchers. The battle noise rose to a climax, as the platoon down at AX went apeshit trying to repel the attack.

The lads knew that I had no air, for I'd given an all-stations warning just as soon as I'd got word of the sandstorm. Via the Predator downlink I could see the enemy pouring in a barrage of fire on to AX. Tracer rounds and RPGs were streaking through the air and slamming into the sandbagged position on the compound roof. With the GeoCell map I could pinpoint exactly where the enemy were hitting us from: it was a position codenamed Golf Bravo Nine Two. If only I had some proper warplanes overhead, I could well and truly smash them. We needed some air power badly, for the platoon down at Alpha Xray looked in danger of being overrun.

I could feel my frustration levels boiling over. But just as quickly as the contact flared up, it died down to almost nothing. As it did so the enemy chatter started going wild.

'We have tested their defences, brothers, and now we know where to hit them,' a Commander Jamali was telling his men. 'Remain in your hidden bunkers and trenches: await the order for the main attack. But if the infidels come out on foot, open fire and kill them.'

It looked as if that was just a skirmish, then. The big battle was still coming. I lost *Overlord Nine Three* – the Predator – which was out of flight time. With the sandstorm raging I had no more aircraft, and the airspace above the Triangle remained empty.

That airspace – my airspace – had the codename ROZ Suzy. ROZ Suzy covered a 7.5-nautical-mile square box of air encompassing Adin Zai, Rahim Kalay and all terrain in the Triangle. I had to declare 'ROZ Suzy hot' if it all went noisy, and 'ROZ Suzy cold' when the contact was over. Right now ROZ Suzy had gone cold, but it was anyone's guess as to how long it would stay that way.

At stand-to that morning we'd discovered a mangy old mutt asleep under the Vector. It was like he'd blown in with the sandstorm. He looked a bit like a Labrador, but he was in a seriously bad way. Every one of his ribs was showing; his fur was hanging off in big clumps; he was full of fleas and mites; and his two front feet were all askew, as if he'd had his legs broken.

I'm a self-confessed dog-lover with two British bulldogs – Trevor and Honey – cluttering up the living room at home. Once the fighting was over, we took the stray to the well and scrubbed him down with British Army hand wash. That dealt with the parasites. He was so weak we thought he was a goner, but we did our best to get some food into him.

Bully beef turned out to be a real favourite, and in no time he was perking up. We named him 'Woofer', after The Worcester and Foresters – 'The Woofers' – one of the regiments recently amalgamated into 2 MERCIAN. Woofer was adopted as the FST's mascot, and he took up residence lying by the open door of the Vector.

From somewhere Sticky found Woofer a genuine metal dog's bowl, and tucked it away under the wagon's rear axle. Woofer quickly developed a liking for sausage and beans, and there never seemed to be any shortage of lads who wanted to unload some on him. Every now and then Woofer would wander off, but he was always back at feeding time.

Over the next twenty-four hours there was a growing sense of impending menace in the Triangle. At Monkey One Echo we had a 'walk-in' – a local elder who came to the base offering up Intel. He told us that the enemy had shipped in two hundred extra fighters, and that they'd resupplied themselves with shedloads of weapons and ammo. We asked him where the enemy were positioned, but he just shook his head. He had no idea exactly. He pointed downhill into the Green Zone.

'Two hundred metres that way,' he said. 'Down there by two hundred metres.'

The enemy were preparing for the 'big push', he told us, to force us out. I was getting more and more frustrated. What were the enemy waiting for? If there was going to be a shitfight, let's get it on was my attitude. But it was often like this in the Army: *hurry up and wait*.

In the burning heat of the afternoon Sticky, Chris, Throp and me decided to have a swim. There was a tributary of the Helmand River some fifty metres into the Green Zone. It was pretty insane to take a dip in the midst of fighting a war, but we needed to do something to defuse the tension. We were going spare with the heat and the waiting, and we needed to cool down.

Somebody had blown up the bridge over the river. The locals had improvised a new one using a gnarled old tree trunk, which was bleached white in the sun. It made the perfect platform from which to take a dive. The river was deep enough, and marvellously refreshing.

A couple of young 2 MERCIAN lads chose to come with us. Sticky stood guard while we larked around diving off the log, and holding on to it with the current roaring past. I noticed the 2 MERCIAN lads having a whisper and a giggle, and then they disappeared upstream. A few minutes later Throp saw something floating down towards us. It looked like a stick. He went to grab it, so we could chuck it around a bit.

'Oh shit!' he yelled. 'It's a shit! Literally it is!'

Throp had grabbed hold of what turned out to be a massive man-log.

'You're fucking joking!' I yelled. 'It's a human land-mine!'

Throp and I were laughing so much we were half-drowning. Up on the riverbank Sticky was killing himself. But Chris wasn't amused. In fact, he had a massive arse on. He stormed out of the river and started yelling in the direction of the 2 MERCIAN lads.

'You fuckers! Fucking get here!'

He was a captain and they were ranks, so they pretty much came running.

'You fuckers!' he roared at them. 'You just did a shit towards us!'

'We didn't do it, sir!' one of the lads protested.

'It wasn't us!' said the other lad.

You could tell in an instant that they were lying, and Throp, Sticky and I were doubled up. The more Chris lost it, the funnier it became. Eventually, he threatened to put the two lads on a charge. It was priceless.

'What are you going to charge them with?' I gasped. 'Defecating in a forbidden position? Or defecation of duty?'

Later, back at Monkey One Echo, the guilty party came to have a quiet word with me. He was looking a bit sheepish.

'Erm ... about what happened in the river ...' he began.

'Shut up, man,' I cut in. 'It was class.'

He grinned. 'It was me who did the turd. But don't tell Chris, eh?'

I promised him that I wouldn't. Just as soon as I saw Chris, I told him who it was had laid the log in the river.

'So mate, what d'you reckon you're going to charge him with?' I needled him.

Chris had calmed down a bit by now. He had to laugh. After all, what was a little shit between blokes, when soon we'd be fighting back-to-back to save each other's lives?

When it all went noisy the only place to get eyes on the Green Zone was sat atop the old boy's back wall. From there I could see everything, but I was also a bullet magnet. As the rounds went flying everyone was running for body armour and helmet, whilst I was getting a leg-up from Sticky on to the wall. The 2 MERCIAN lads would be staring at me like I was insane. To be honest, I never used to think about getting shot. If I did, I knew I'd never go up there. And it was the only place where I could see the battlefield, to talk in the air.

With the enemy chatter buzzing that they were poised to attack, the OC decided to push a patrol down to Alpha Xray. He wanted to check on the platoon in the Alamo, and to recce around their position. We formed up with the lads from Monkey One Echo, in a convoy of WMIK and Snatch Land Rovers, with Throp and me on a quad bike.

The quad was a beast of a thing, with four knobbly wheels like mini-tractor tyres. It was mostly for Sergent Major Peach to use, as a fast and manoeuvrable ammo resupply and casualty evacuation vehicle. I'm a bit of a petrol-head, and no way was I letting Throp drive. I stuck him on the pillion and grabbed the controls.

We left Monkey One Echo at 0900 and were out patrolling the Green Zone for four hours. The OC hoped to provoke the enemy into showing their hand. We were frustrated and bored and dying to get a reaction, but there was nothing. By the time we turned up the dirt track that led back to our base, not a shot had been fired at us.

Halfway up the hill I pulled a massive wheelie, without warning Throp. In an instant I had the beast rearing up on its back wheels, with Throp screaming that he was falling off. I held it like that for several seconds and then slammed it back down again, by which time the two of us were laughing our tits off. But not for long.

I glanced ahead to check for the convoy, and realised that we'd lost them. At that very moment the enemy opened up. There was an ear-splitting burst of gunfire, and the first rounds slammed into the vegetation barely metres behind us. I twisted the throttle with sweating hand, giving it maximum revs, and went flying up the track like a bolt from hell. I reckoned the convoy had to be just ahead of us.

As we bucked and weaved our way through the trees, more and more weapons were opening fire. Bullets hammered into the dirt throwing rocks and shit all over us. Over the roar of the quad's engine I could hear something horribly big hammering away – thud-thud-thud-thud-thud – the rounds from which were chewing up the bush. It sounded like a Dushka, and it was absolutely fucking terrifying. A Dushka round would take your head clean off, or rip your arm or leg from your body. All it would take was one 12.7mm bullet tearing into the quad, and it'd be smashed to pieces. Or a round in the petrol tank, and Throp and I would be nicely torched.

Our only hope lay in speed. It takes great skill to shoot and hit a man sprinting at speed, or two lunatics on a careering quad bike. I drove the race of my life, throwing the machine from side to side and cannoning off the banks to either edge of the track, and flying over ruts and bumps, ducking branches and vines as we went.

We slewed around a blind corner, and up ahead were the vehicles. I gave it one final blip of throttle, caught up with the rear WMIK, and grabbed one of the vehicle's handrails. We had them tow us into Monkey One Echo, under the safety of their 50-cals and

Gimpys. I powered down the quad and sat astride her for a second, my whole body shaking. Whether it was from nerves or the bone-rattling ride I wasn't sure. I'd been shitting myself out there. As for Throp, he looked as white as a bloody sheet.

Now we knew just how closely the enemy were watching us, and how many of the fuckers were out there. The moment they'd got us on our own, they'd unleashed a storm of lead on us.

It promised to be some shitfight, whenever it might come.

SIXTEEN
THE SIEGE OF
ALPHA XRAY

We moved up to PB Sandford using Route Buzzard, a dirt track that led along the ridge line. Accommodation here was five-star, compared to where we'd just been. Sticky and I grabbed our own room, in what must have been the camel or donkey house.

There was the one arched doorway with no door, a dirt floor, and a couple of mud-sculpted feeding troughs for whatever animals had once lived here. I managed to scrounge a camp bed, which was sheer luxury. Sticky cobbled together a bed of sorts from an old frame made of branches and strung with paracord for a mattress. It was a stickyback-plastic-and-cardboard-box kind of bed, and Sticky bitched a lot 'cause I had a real camp bed. I told him that I had a camp bed because I was the JTAC, and the JTAC needed a good night's kip. As it happened, it was too hot to sleep in that room during the day, and too stuffy at night. We'd have been better off dossing down outside.

The roof above was a dome of smooth mud plaster, and the topside became my new JTAC position. Being domed, it provided an all-around view south over the Green Zone, and north over the desert. It was perfect for controlling warplanes, but provided zero cover from enemy fire.

Along with the move to PB Sandford, we'd picked up a fifth member of the FST. Bombardier Karl 'Jess' Jessop was going to be

Chris's right-hand man, cueing up the field guns, and leaving Chris free to keep an overview of FST operations.

Chris played hockey for England, and Sticky, Throp and I were always winding him up about it.

'Hockey – that's a girls' sport, isn't it?'

'No it's not,' Chris would argue. 'It's for real men.'

It turned out that Jess was also a keen hockey player, so we reckoned the two of them were made for each other. But it soon became clear that there was this love-hate thing between them. Maybe it was about who was the better hockey player? Fuck knows. Either way, the rest of us were determined to get some good mileage out of it.

It was a suffocatingly hot afternoon and I'd gone to get a brew on. The chill-out area at PB Sandford was another real luxury. From somewhere the lads had got a massive, soot-stained tin kettle. It sat on a clay stove, with a fire burning whatever wood we could find beneath it. You could get a brew on at any time of day or night, and there was a U-shaped arrangement of 'sofas' cobbled together from wire mesh and sacking.

Whilst I'd been away getting my AIDS test some cheeky bastard had stolen my mug, so I'd blagged a new Jack flask from stores, and stuck a big square of white gaffer tape on the side of it. On that I'd scrawled in black marker pen: FAT JTAC'S MUG – TOUCH THIS AND I'LL JDAM YOU. It seemed to be working, for whenever I went for a brew my Jack flask was waiting for me.

I was supping my tea sweet and strong, just as I like it, and marvelling at how it could be so suffocatingly, crushingly hot. The heat would build through the afternoon, until it was a furnace inside and out. The one big downside to PB Sandford was there was no river to take a dip. There was just the one, bloodworm-infested well in the centre of the compound. Come mid-afternoon there was a queue of overheating British squaddies seeking a bucket-load of water to chuck over their heads.

Above the natter of the lads I heard a faint boom. It sounded as if it had come from the south-east of us, in the direction of Qada Kalay. The longer-term plan was to get a fourth patrol base (PB) established, on the southern side of the river, to choke off both sides of the Green Zone. Qada Kalay was where we intended to build it, and it was a known enemy stalking ground.

I raced outside, mug of tea in hand, and pounded up the steps that led to the rooftop. As I got there, I heard a familiar sound – the faint whistle of an incoming 107mm warhead. We had a rocket inbound. Mikey Wallis, the dude with his mortar-locating kit, was with us at PB Sandford, but for some reason his gizmo didn't always register 107s.

'107 INBOUND!' I yelled.

As the lads ran for cover the warhead swooped down on us, howling like a banshee. It ripped into the desert right on our south-eastern wall. From up on the roof I saw the blinding oily-yellow flash of the explosion, and the shockwave of the blast tore over me. As the ear-splitting roar of the detonation echoed around the Green Zone, a gout of angry smoke fisted darkly into the sky. The warhead had landed fifteen metres outside our front wall, and it had been fired from a kilometre and a half away. Not bad for a first shot.

I needed to get air over Qada Kalay pronto, so I could find the launcher and smash it. I dialled up Widow TOC, and as luck would have it there was a Predator barely a minute out from my ROZ. At least it would enable me to get eyes-on. I radioed the operator.

'*Overlord Two One, Widow Seven Nine.* Sitrep: we're under a 107mm rocket barrage, being fired upon from 1.5 kilometres south-east. I want you over that position looking for a tripod launcher tube.'

'Roger that, sir,' came back the reply. 'Banking around now.'

Positions on the opposite side of the Helmand River had been allocated November codenames, and I gave the operator the GeoCell location of where I thought the 107mm had come from: November Nine Five.

There was always a weird disconnect when controlling a Predator. Here was I, a British JTAC under a burning Afghan sun, talking to a 'pilot' who was flying the thing remotely from an air-conditioned bunker in Nevada. He'd be sat in some padded leather chair staring into his screen, whilst I crouched on a mud roof under a plume of smoke from a 107mm rocket strike.

However much the American operator might try to sound as if he was 'with' me, we both knew he was a million miles away. More than likely, he'd stopped off at a Hooters for a mountain of dough-nuts and a gallon of coffee en route to work, and was looking forward to a few chilled beers once his shift was done. Getting a Predator operator's head into the mindset of the war we were fighting was never easy.

Sure enough, the operator did discover a metal A-frame at the November Nine Five location. I took a good long look at the grainy image on the Rover screen. Like most feeds from 20,000 feet, the Predator's was jumpy and prone to breaking up whenever the aircraft hit turbulence.

I wasn't sure about that A-frame. I called one of the 2 MERCIAN lads to take a look – a guy who was a weapons boffin. He studied the image for a few seconds, then shook his head. There was no way it was the tripod for a 107mm launcher. It was too widely splayed apart for that. It was more than likely a frame to hang a cooking pot over an open fire.

I had the Predator for several hours, during which time a couple more 107mm rockets slammed into the base. We combed the terrain all around Qada Kalay, but nothing. No one was injured from the 107mm strikes, but the enemy chatter was going crazy. Their leader, Commander Jamali, was crowing about the rockets scoring 'direct hits' on us lot.

Jamali called for his men to gather at 'the mosque' for a pre-attack briefing. *Come on then*, I was thinking. *If you're going to hit us, fucking hit us. Let's get it on.* I guess the OC must have been feeling equally frustrated, for that evening he issued a set of orders to take the fight to the enemy.

At 0300 we were to push a stick of lads into the Green Zone on foot. It was to be a fighting patrol at night, and the aim was to root out the enemy. Via Routes Buzzard, Sparrow and Crow – three dirt tracks threaded through the bush – we'd do a U-shaped circuit of the Green Zone, the furthest extent of which would be Alpha Xray.

It was to be a platoon-strength patrol – so twenty-odd men – and Sticky and I were on it. As the two of us prepared our kit, stuffing spare mags and water and grenades into our packs, I heard Chris let out a laugh.

'What the fuck is that?' he demanded.

I glanced up and there was Chris pointing at Sticky's Bergen. Sticky had one of those little fluffy Snoopy dog key rings hanging off one of the straps.

Sticky grinned. 'It's a Snoopy key ring.'

'I know that,' Chris snorted. 'I mean, what's it doing on your backpack?'

'It's a present from my girlfriend,' said Sticky. 'Kind of like a lucky charm.'

'Well it looks chippy as fuck,' said Chris. 'Get it off.'

'Can't,' Sticky said. 'It's my good-luck charm. I promised her I'd wear it.'

'Look, we're meant to be a bunch of professionals attached to an infantry company,' said Chris. 'But you've got a cuddly toy swinging from your kit. I'm not happy with it.'

Sticky shrugged. 'Yeah, OK. I'll take it off.'

By the time we'd mustered for the patrol at 0245, the Snoopy dog was still swinging from Sticky's pack. You had to laugh.

The OC had warned us to expect danger-close engagements, and I had air booked for the duration of the patrol. I had a B-1B supersonic bomber, call sign *Bone One Two*, ramped up high. And below him I had a lone Apache, *Ugly Five Zero*. I tasked the B-1B to search for any vehicles moving in the Triangle, in case the enemy bussed in reinforcement to hit us. I tasked the Apache to fly directly overhead, shadowing the patrol. When the enemy showed themselves I'd give the pilot a simple direction and a distance from us, and he'd mallet that position.

There was a quiet, nervous tension as the lads gathered in the muggy darkness. It was boiler-room hot, even at this hour. It was especially sticky, what with all the kit we were carrying. The 2 MERCIAN lads were mostly in their late teens or early twenties. As they snapped their night-vision monocles down over one eye, I sensed a hunger to get out there and get in amongst them.

The gates to the base creaked open. The atmosphere was electric. We were about to venture into the dense bush of the Green Zone, on foot and in pitch darkness, knowing the enemy were all around us. We'd had two good men killed fighting for control of this territory, and a dozen or more injured. We were about to walk into the fire once again.

We filed past the front sangar – the sandbagged position at the gate – threaded our way down the escarpment and into the wall of darkness. No sooner had we hit the vegetation, than the air traffic started going mad.

We had Naji, our regular terp, with us. Naji was a quiet but friendly guy, with a shyness about the eyes. He'd had half his family murdered by the Taliban, and he hated the Talibs with a burning vengeance.

As soon as the enemy got on the air, Naji started translating. 'Get up! Wake up! They're coming in on foot! The Diamond Special Forces are coming!'

I felt a shiver half of fear and half of excitement running up my spine. Apart from the scrunch of boot soles on gravel, and the suck and blow of our breathing, we weren't making the slightest noise. And apart from the faint fluorescent glow thrown off by each soldier's monocle, we were invisible to the naked eye. The enemy had to have night-vision. That was the only way they could know that we were coming.

Diamond Special Forces – that was how the enemy referred to the 2 MERCIAN lads. The B Company boys sported a distinctive shoulder badge – a triangle of green over a triangle of red, making a diamond shape. It was their regimental flash, one designed to enable rapid visual recognition, and it had to be why the enemy had named them the 'Diamond Special Forces'.

If the Taliban had night-vision, the only real advantage we had over them was the air cover. I scanned the wall of undergrowth to either side of us. You could hide a bloody army in there. By all accounts the enemy had. And even with night-vision, you could stumble right over a well-hidden adversary before you spotted him.

We were halfway to Alpha Xray and still Ugly had detected nothing. With its state-of-the-art thermal-imaging systems the Apache could detect a mouse farting at two thousand metres. But apart from the twenty-odd soldiers of our patrol, not a thing could be seen moving.

We called a halt. The night-dark silence closed in on us, predatory and menacing. Sticky dropped to one knee right on my shoulder, his SA80 levelled at the wall of darkened vegetation. One of the 2 MERCIAN lads provided cover the other direction, with a string of lads to the front and the rear. I grabbed my TACSAT. It was time to try something different. It was time to flush the bastards out of hiding.

'*Ugly Five Zero, Widow Seven Nine.*' My whisper sounded like a scream in the crushing stillness. 'I'm bringing the Bone call sign in. Move off to the south of the River Helmand, until advised otherwise.'

'Roger,' came back the Apache pilot's reply. 'Moving south of the river now.'

The chthwoop-chthwoop of the Apache's rotor blades faded away above us. I dialled up the B-1B.

'*Bone One Two, Widow Seven Nine.* I want you to fly a show of force with flares over our position, and all up the Green Zone.'

'Affirm,' came the pilot's reply. 'Banking around now. I'm coming in at 2,500 feet. Stand by.'

I smiled to myself. The massive stealth bomber would be coming down to around the same altitude as the one that had buzzed the camp commandant's briefing, at FOB Price. If anything was likely to get the enemy on their feet and moving, this was it.

'Running in now,' the pilot announced.

A few seconds later there was a roar like a tidal wave sweeping down the valley. For an instant this massive deltoid shape loomed out of the dark sky, silhouetted against the stars. And then it flashed past above, tearing apart the darkness and the silence with an ear-shattering violence. As the echoes crashed over the Green Zone, the pilot fired off a trail of blinding flares in his wake. Every soldier held his breath, as we waited for the enemy to react, or to show themselves. But as the echoes rolled away, not a thing could we see moving out there.

I got the Apache back over us right away, but the Green Zone was a dead zone as far as the pilot could ascertain. A blanket of silence had descended over the terrain once again. But we knew the enemy were out there, and spitting-distance close. The air traffic was going wild with calls that they were visible with us. The discipline of the Taliban was incredible. It was spooky. It had me spooked, any road. They could see us. They knew we were here. But they were holding their fire.

I got the B-1B down even lower, flying a second show of force, but not a sausage. By 0530 we were back at the gates of PB

Sandford. Not a round had been fired at us, and we'd seen not a sign of the enemy.

Back inside the base we reflected on what we'd learned. The enemy clearly had a plan of attack, and they were sticking to it. They were going to take us on at the time and place of their choosing. But like all good plans, theirs had to have a weakness, if only we could find it.

I got my bracket down and managed to doze until around midday. I woke pooled in a slick of my own sweat. I was sleeping beneath a mozzie net, one that was sown into the camp bed to make a kind of pod. It was as hot and breathless as an oven. I struggled out, went to the well and doused myself awake. Time to get a brew on.

At 1530 I got allocated air. I got a Dutch F-15, *Rammit Six Three*, flying recces over the Green Zone. The Dutch F-15s had no Rover downlink, but I had a pretty good view of things from up on my rooftop position – JTAC Central. The pilot had been flying search transects for twenty minutes or so when I got the call.

'Visual with build-up of males of fighting age at Golf Bravo Nine One,' the pilot told me. 'Visual with male pax to the north-west of there, one-twenty metres from Alpha Xray at Golf Bravo Nine Zero. Male pax appear to have heard me, and are disappearing into the treelines. No weapons visible.'

I had the GeoCell map spread out on the roof and weighted down with my fag packet and lighter. We had one group at Golf Bravo Nine One, three hundred metres due east of Alpha Xray. Another, hidden group was a hundred and twenty metres away from our lads. At the same time we were picking up radio chatter about the enemy being ready to attack.

What did it all mean? The enemy had been harping on about being poised to attack us for days now, yet nothing much had happened. I passed what I'd learned from the F-15 to the OC, then lodged it away in the old grey matter.

At last light I lost the F-15, which was low on fuel. Nothing more had been spotted, and all was quiet in the Green Zone. I was down by the Vector having a quiet smoke, and wondering what the hell the enemy were up to, when an almighty explosion rocketed across the Green Zone. It came from the direction of Alpha Xray.

An instant later there were repeated, deafening explosions, followed by machine guns opening up on the base in a solid wall of sound. Almost immediately, there was the scything roar of the Gimpys on the rooftop position returning fire, joined by the mauling thump-thump-thump of the 50-cals.

Alpha Xray was under siege, and meeting fire with fire. This was the big one. This was what we'd been waiting for. I got on the TACSAT screaming for air.

'*Widow TOC*, *Widow Seven Nine*,' I yelled above the battle noise. 'Sitrep: troops in massive contact. Requesting immediate CAS.'

'Roger. Stand by.'

'Sticky, get a sitrep from AX,' I yelled at him.

'The enemy are hitting AX from four fire points,' Sticky relayed the update from the platoon commander. 'They're taking fire from all directions.'

Via Sticky, I got the platoon commander to describe those fire points. From what he told us, the enemy had to be in the woodline that the F-15 pilot had spotted earlier. They'd made a fatal mistake. They'd shown themselves too early, and revealed their location to our eye in the sky. Now I knew where to hit them.

'*Widow Seven Nine*, *Dude Zero Five*: inbound into your ROZ two minutes.'

I had an F-15 powering in to the battle space. I radioed the pilot.

'*Dude Zero Five*, *Widow Seven Nine*. Sitrep: troops in heavy contact at Alpha Xray. I'll AO update once we've finished the attacks. Confirm you're happy.'

I didn't have the time to bugger about talking the guy around the battlefield. All I wanted to do was pass him the GeoCell position, and get him smashing the enemy.

'Roger,' the pilot confirmed. 'Happy with that. Just tell me where you want me doing the drops.'

'First target is W-shaped treeline running north of Golf Bravo Nine Zero,' I told him. 'Enemy are one-twenty metres danger-close to friendlies at Alpha Xray.'

'Affirm: visual with treeline at Golf Bravo Nine Zero,' the pilot replied. There was a short pause. 'Visual with muzzle flashes plus three heat spots running around in treeline.'

'Roger – stand by. Chris,' I yelled. 'Tell the OC I'm going to smash enemy positions in the woods at Golf Bravo Nine Zero, then Golf Bravo Nine One.'

Chris relayed the message to the OC, and Butsy said to hit them. The roar of the firefight was building to a climax, and I had to get the jet in. But with the airstrikes going in 120 metres, there was no way I was about to start dropping bombs.

'*Dude Zero Five*, you're clear to attack. I want you to hit the target on a south-west to north-east attack run, with a 20mm strafe. Nearest friendlies one-twenty metres west of target.'

'Roger – ninety seconds out,' came the pilot's reply. A beat. 'I'm visual more pax running around a compound, and firing from rooftop positions.'

'That's our compound!' I yelled at him. 'That's us! That's friendlies!'

'Roger that,' the pilot confirmed. 'Visual friendlies. Call for clearance.'

'Chris – sixty seconds for cannon!' I yelled.

Chris gave me a thumbs-up, and bent to his radio to pass a warning to all stations that the strafe was coming in. For a moment I considered running up on to the roof to get visual, but I knew there wasn't time.

'*Dude Zero Five, Widow Seven Nine*: I'm not visual your attack. Repeat: not visual. You're clear hot.'

'In hot,' the pilot confirmed. 'Engaging.'

'Brrrrrrrrrrrrrrrrrrrrrrrrrrrrrrrrrrrrrrrrrrrrrrrrrrrrrrrrrrrrrrrrttttttttttttttt.'

For what seemed like an age the deep-throated growl of the strafe hammered around the walls of PB Sandford. I wished to hell I'd seen it go in. I needed to know he'd nailed the enemy and not hit our lads. Knowing that I'd sent him in blind – without eyes-on – the pilot came back with an immediate BDA.

'BDA: I put a strafe three hundred metres down that treeline. Two pax confirmed dead.'

'Roger. Good work, *Dude Zero Five*. Bank around south, and do an immediate re-attack on new woodline. Stand by for grid.'

As I pawed the map, trying to work out the eight-figure grid for the next hit, the noise of battle was loud as ever. The next enemy position wasn't identified by any Golf Bravo prefix on the map, so I needed to talk the pilot on and lock him on to a grid.

'Next target is a south-east to north-west treeline. Bisected halfway by a shorter treeline at right angles, forming a X-shape. Grid is: 03759284. Readback.'

The pilot confirmed the grid, and slaved his sniper optics to the coordinates.

'Visual X-shaped treeline,' the pilot reported. 'Visual two males lain at the base of trees, with muzzle flashes.'

I ordered him to attack, and cleared him in to do the strafe. The F-15 put a long burst of 20mm cannon fire into that second woodline, tearing the position to shreds. The pilot reported immediately that two more enemy were dead.

Still there was a barrage of RPGs and machine-gun fire slamming into Alpha Xray. I got the pilot to bank around north and do an immediate attack on the third enemy position. I passed him the next grid, and gave him the talk-on.

'Visual with three pax running up the east side of that woodland,' the pilot reported. 'They're running away from your friendlies. Tipping in.'

If the enemy were running, maybe the airstrikes were breaking their will to fight. The pilot put a long strafe into that wood, his third in as many minutes.

'BDA: one killed,' the pilot reported. 'Plus I can see one pax dragging a wounded figure by his arms to the east.'

'Roger: leave them,' I told him. 'They're out of action. Scan the woodlines to the north of Golf Bravo Nine Zero.'

'Roger. Scanning now.' A short pause. 'Visual with two pax with weapons on their backs crawling towards your friendly position.'

'Attack from west any way you can to the east,' I radioed the pilot. 'Keep the strafe away from friendlies.'

The F-15 came screaming in on its fourth attack run. The jet's six-barrel cannon roared, saturating the woodline with high-explosive 20mm rounds.

'BDA: two more dead,' the pilot radioed. 'Low fuel. Bugging out. Stay safe, *Widow Seven Nine*. Out.'

As the F-15 left the battle space, the contact down at Alpha Xray was still rumbling and smoking. Just as soon as the jet was gone, the crack of gunfire and the ripple of explosions spiked. The enemy fighters must have realised that we had no air cover.

Again, I was back on the TACSAT screaming for jets. I got *Recoil Five Five*, a Harrier, inbound four minutes. We'd improvised a staircase out of ammo boxes leading up to my rooftop position. I raced up to get eyes-on.

From JTAC Central, I could see the sparking of muzzle flashes and the flaming kickback of RPGs. In spite of the repeated strafes, the enemy had Alpha Xray surrounded and were closing in. Where the fuck was that Recoil call sign?

From below me Naji, our terp, started yelling out some intercepts of enemy comms. Commander Jamali was screaming for his men to press home their assault. With the skies above the battle clear of warplanes, they were to overrun their objective – Alpha Xray.

'I'm in my hardened position!' Jamali kept yelling to his men. 'I'm safe in my hardened position! Push onwards with the attack! Overwhelm them!'

I was standing on the domed roof scanning the terrain below. Where the fuck was this Jamali's bunker – his 'hardened position'? A round cracked past, whining off the mud roof. I guessed an enemy sniper was on to me. But the light was fading fast, and I didn't rate his chances. In any case, I wasn't moving. I ran my eyes across the Green Zone. Where was this Commander Jamali? Where was this bunker? Where in the Green Zone could his 'hardened position' be? If I searched hard enough, might I be able to sniff this Jamali out?

As it happened, I was just a whisker away from nailing him.

SEVENTEEN
TEA AND CRICKET

Major Butt and Chris joined me at JTAC Central. We had a few hurried words about this Commander Jamali fella. He was clearly the big cheese in the area. We threw around a few ideas about where his 'bunker' might be, but we didn't come up with anything definite.

The Harrier checked in to my ROZ. I got him banked up to 25,000 feet, so the enemy couldn't hear him. I passed him the search coordinates and got him scanning for enemy RPG or small arms firing points. As the pilot began his search, I flipped out my Rover screen and logged on to the downlink. The Harrier had some awesome avionics and night-vision capabilities. The terrain below the jet appeared on my screen in close-up, ghostly green detail. The hotter a heat source – a human form; a warm car engine; a recently fired gun barrel – the more it showed as a glowing shape picked out in fluorescent green.

Chris and Butsy gathered round, our eyes glued to the grainy image. The battle was still raging, our lads and the enemy trading fire with fire. Tracer arced through the darkened sky, painting angry red lines across the valley. We just needed the Harrier to find those firing points.

At 2015 I got the call.

'*Widow Seven Nine, Recoil Five Five*. Visual RPG position north of the treeline at Golf Bravo Nine Two. Visual armed enemy pax on a compound roof at that position.'

As he said the words, there was a flash of green like a water splash on the Rover screen – the blast of an RPG being unleashed at our lads. The glow of the rocket firing lit up the entire enemy position. On the south-west corner of the roof there was what looked like a sangar. Nine heat spots – human-shaped ones – were lying in and around it.

'*Recoil Five Five*, *Widow Seven Nine*. I want you to hit that position with a thousand-pound JDAM.' As I said the words I glanced at Butsy, who gave me the nod. 'I want you in on a south to north run, and I want the bomb put through the roof of that building.'

'Roger. Thousand-pound JDAM on a south to north run. Positioning. Stand by.'

'Nearest friendlies three hundred metres to south-west. Call for clearance.'

'Tipping in.'

As the Harrier pilot began his attack run, the figures on the Rover screen ceased firing. We watched the glowing blobs grab their weapons, and disappear through a door into the building. They must have heard the jet overhead, and they were taking cover.

'Sixty seconds out,' came the pilot's voice. 'Call for clearance.'

'No change friendlies,' I replied. 'I'm not visual your attack. Repeat: not visual. Clear hot.'

'In hot,' the pilot radioed. 'Stores.'

In the JDAM came, a low whistle from the direction of the Helmand River, rising over several seconds to a horrible, howling scream. It sounded like nothing else on this earth. It was like a B-1B pilot had gone kamikaze, and was flying his giant, supersonic bomber on a suicide mission into the heart of the Green Zone.

As it hit, there was the violent, white-hot flash of the detonation, and the Rover screen broke up into a thousand shards of light. I lifted my head from the terminal, and the massive roar of the blast hit us. In the heart of the darkened bowl a fountain of fire erupted. It was like a volcano was vomiting red-hot lava and smoke into the night sky, flinging out burning rock and debris far and wide. For several seconds the entire scene was lit up an unearthly red, as the explosion plumed and boiled. Woodstrips, ridges, the trees lining Routes Crow and Buzzard – all were picked out in angry silhouette, giving me a rough idea where the JDAM had hit. I radioed the Harrier. It looked as if the strike was bang on target, but I had to be sure.

'*Recoil Five Five, Widow Seven Nine*. BDA.'

'BDA: direct hit on compound roof. Enemy position destroyed.'

As the explosion died down to a scatter of angry fires, the image on the Rover screen stabilised. All that remained of the target was a smear of shattered rubble scattered in a funnel some five hundred metres north-east of the JDAM's impact point. The attack run had thrown the blast and the debris away from Alpha Xray, just as I'd intended.

There was a cry from Naji, our terp. The radio chatter was going crazy. The enemy were screaming for 'Commander Jamali' to check in, but no voice was answering. All the enemy were getting in response to their calls was an echoing void of silence.

I locked eyes with Chris, and the OC. 'Fucking hell. Are you thinking what I'm thinking? D'you reckon we got him?'

'Could be,' Butsy remarked. 'Maybe that was his "hardened bunker".'

I nodded. 'Not hard enough for a JDAM, though, eh?'

The battlefield had fallen strangely, eerily silent. The Harrier had come to my aid from another firefight and was low on fuel. I got on the TACSAT and requested air, and I got an A-10 Warthog ripped to me from another contact. He was fifteen minutes out.

As we waited for the jet, the platoon commander at Alpha Xray reported that they could see figures in the treelines with torches. The enemy were out searching for their injured and their dead. I got the call from the Warthog and gave him an AO update.

'*Hog One One*, *Widow Seven Nine*: we've got enemy pax in the treelines all around our position at Alpha Xray. They're out with torches picking up casualties. They're not in contact, so do not engage. Repeat: do not engage.'

I talked the pilot around all the contact points, and then he had this for me: 'Visual with two enemy pax crawling through the tree-line from the site of your JDAM strike towards your friendlies. One has an AK-47 on his back, one has an RPG.'

The A-10 had no downlink capability, so I had nothing on the Rover screen. But they were clearly moving into positions where they could re-attack Alpha Xray, and they were armed: that made them fair game. The OC told me to hit them.

'*Hog One One*, *Widow Seven Nine*: I want you to attack using CRV7 rockets, on a north to south attack run. Friendlies two-fifty metres to the south-west.'

'Roger. Two minutes out.'

I couldn't ask for a strafe, for the A-10's 30mm Gatling gun would tear up the entire treeline, and I knew the enemy were out collecting their dead and wounded. The CRV7s would make a direct, targeted strike, and were as big a hit as I was willing to risk danger-close to friendlies.

I cleared the A-10 in, the pilot diving hard and fast, aiming his jet and the rockets directly on to target. The flash of the CRV7s flared beneath the A-10's wings, then: boom–boom! The double-crack of the rocket's impact rumbled through the darkness.

'BDA: direct hit on one of 'em,' the pilot reported. 'The other's set off like a bat out of hell. I lost him.'

I got the A-10 flying low and noisy orbits over Alpha Xray for the next twenty minutes. Nothing else was seen. *Hog One One* was ripped by *Hog One Five*, but the Green Zone was totally dead by now. At 2215 *Hog One Five* had to bug out, low on fuel.

I had no more air, it was all quiet down at Alpha Xray, so we decided to get the kettle on. Chris, Throp and I sat having a brew, with our head torches casting a warm glow over the chill-out area. I'd hardly wetted my lips when Sticky yelled over that I was wanted on the radio. I had an Ugly call sign trying to raise me on the air.

I grabbed my TACSAT and headed on to the roof, being careful not to spill my cuppa.

'Ugly call sign, this is *Widow Seven Nine*.'

'*Widow Seven Nine*, *Ugly Five Two*. I'm en route back from Sangin, and I can offer you thirty minutes' playtime. I hear you've been busy down there.'

'Fuck, aye we have.' I gave him an AO update, being sure to mention the JDAM strike. 'I want you flying air recces over all points of recent contact around Alpha Xray.'

'Roger. Commencing air recces now.' As the pilot got the Apache's nose-pod scanning, the two of us got chatting. 'So how's it been down there? Good?'

'Aye. It's been a top job,' I told him. 'What's the graft down at Sangin?'

'Not a lot. Sounds like you're getting all the action up here. Hold on, I've seen something. Stand by.' A beat. 'I'm visual three pax hiding in the woods directly to the east of Golf Bravo Nine Zero. They've got eyes on your friendlies at Alpha Xray.'

'Can you see any weapons?' I asked.

'Negative.'

'Keep an eye on them.'

For ten minutes the Apache kept his pod zoomed in on those three males of fighting age. I had no downlink from the Apache, but the pilot was giving me a running commentary.

'Visual those three pax, two of whom are now showing weapons.'

'Hit them with 30mm,' I told him. 'Nearest friendlies Alpha Xray two hundred metres south-west.'

'Roger. Engaging.'

Thump-thump-thump … The dark heartbeat of the Apache's cannon rumbled out across the night, as the single-barrelled 30mm gun spat out a ten-round burst.

'Two direct hits,' came the pilot's voice. 'The third guy's on his heels with a weapon running for cover. Engaging.'

The Apache spat fire again, a second ten-round burst of 30mm cannon fire chasing the lone fighter up the treeline. The pilot played cat-and-mouse with him for a further two, ten-round bursts, before finally he ceased firing.

'Third enemy killed,' the pilot confirmed. '*Widow Seven Nine*, I'm low on fuel and returning to base. You stay safe down there.'

The rotor blades faded away on the warm night air. I glanced at the dial of my watch. It was just past midnight. I'd been controlling jets for little short of twenty-four hours, with just a few hours' kip in between. As the adrenaline drained out of my system, I realised that I was dog-tired.

I was also starving hungry. I couldn't remember when I'd last eaten. I stumbled down from the roof, ripped the top off a sausage-and-beans meal and spooned the lot down. It was lukewarm and gloopy, but it sure as hell did the job. I rinsed my spoon off in my cold tea, glugged it down, and headed for Sticky and my 'bedroom'. I hit the camp bed fully clothed and was out like a light.

The following morning we had an after-action briefing. Between the airstrikes and the fire put down by the lads at Alpha Xray, we reckoned the enemy had suffered serious losses. The jets and the Apache had accounted for fifteen confirmed kills, not to mention the unseen and the uncounted. And the lads at AX had smashed a shedload more.

The enemy were still going spare, calling for Commander Jamali to check in. He wasn't responding. A little later that morning we had a walk-in, a local elder. He was dressed in a flowing white robe topped off with an orange-beaded skull cap, and he had real attitude.

His finger stabbed the air as he jabbered away excitedly to Naji, our terp.

I saw Naji smile. 'He's saying that Commander Jamali was killed last night by the big bomb. The twelve men on the rooftop who died with him were the bodyguard team. Jamali was the top enemy commander in the area. A very important enemy leader.'

Result. More than likely, Commander Jamali was the guy who'd ordered those ten Afghan policemen to be gunned down in cold blood, during our initial assault on Adin Zai. Looks like we'd evened up the score a little. There was nothing for it but to go get a good fry-on.

The last resupply convoy had brought in a stack of sausages and bacon from FOB Price. It was vacuum-packed, so it was still mostly edible. From somewhere Sticky had scavenged an enormous iron flying pan that was stained coal-black from long years of use. He hoofed the kettle off the stove, got the frying pan on, and threw in a big lump of lard. He got the fire stoked up with some scrap wood and hexamine, the Army-issue fuel blocks. And in no time the wonderful smell of frying sausage and bacon was drifting across the compound.

The funniest thing was how the terps seemed to love the fry-ups just as much as we did. Naji couldn't get enough of Sticky's burned sausages. Maybe he kidded himself that they were beef, or something. I reckoned we could get Naji on the ale, if only we had some.

None of us could get Naji's name right. In full, it was Naquibullah, which was a bit of a mouthful for us. So we'd nicknamed him 'Alan'. I guess we chose the name Alan 'cause it was about the most boring-sounding English name we could think of. Apart from Brian, and none of us were cruel enough to call our terp Brian.

Anyway, Naji didn't seem to mind being called Alan, so that became his name. Sticky and him were the best of buddies, and Sticky liked to think of himself as being matey-matey with all the terps. Sticky also reckoned himself a bit of a Pashto speaker. He'd natter on in what he thought was their language, but the terps would just stare at him blankly – which was the perfect opportunity to rip the piss about what a load of bollocks he was talking.

Most of the terps were a great crack. Even those who didn't tuck in to the fry-ups would get some local bread, dip it in the leftover lard and pork fat, and tuck in. But there was one terp, a real loner, who just didn't seem to fit in. He was a shifty bastard if ever there was one. We decided to keep a close eye on him.

The fry-up was followed by a big cricket-off. One of the few things I had done well at when at school was cricket. Coming from the north-east of England, my Redcar school was big into the game, and I'd been a bit of a star batsman.

Chris was another big cricket fan, and he was a fine all-rounder. Sticky had never played before, but it didn't take him long to get in to it. As for Throp, he was a big, well-built lad and he could proper slug the ball. We'd have fifteen to twenty fielders scattered around the compound, and whenever Throp was batting they'd move right back to the HESCO walling.

But it was Butsy who was the real cricket-head. The OC had played at a reasonable level, so he was a big competitive dad. The rivalry between the three of us 'oldies' – Butsy, Chris and I – was far fiercer than anything with the young 2 MERCIAN lads.

Butsy would be out there in his camo shorts and flip-flops, just like the rest of us, and whenever he was playing he was pretty much sure to win. But if the OC was otherwise occupied, it was usually between Chris and me.

Somehow, a proper cricket bat had made it out to PB Sanford. It probably came to Helmand in a Help for Heroes parcel, and got shipped to the base on a resupply convoy. We'd spray-painted some cricket stumps on one wall, and stuck three tubes full of mortar rounds into the dirt, to form the other wicket. We'd cobbled together a rock-hard ball made of rags wrapped round and round with black nasty. The most fun was to be had hurling that ball at the bare legs or torso of whoever was in to bat. If you got hit, it really did chafe.

We'd scratched a line in the sand around the perimeter of the compound, which was the boundary. If your ball rolled over that it was a four; if you whacked it clean over, it was a six. We'd have two innings, and everyone would get two bats. And after the first innings we'd have a break, so everyone could get a brew on.

We'd play for hours on end until there was a contact, and then we'd run around like lunatics getting in position to mallet the enemy. The terps used to stare and stare whenever we were having a big cricket-off. It was like we were insane or something. Whilst they were up for having a laugh, the terps just did not get cricket.

I had no air that day, and so I proceeded to have a cracking good time in bat. Eventually the burning sun drove the lot of us over to the well for a good dousing. After that I made for the bedroom, got my bracket down and tried to doze through the heat of the afternoon. Eventually, I drifted into a deep sleep.

Some time later I shot bolt upright, my world exploding all around me. My heart was thumping like a jackhammer, and my eyes were like saucers. I was shitting myself. There'd been a massive blast right where I was lying, and my ears were still ringing from the boom.

It sounded as if the base security had been breached, and the enemy were chucking grenades in amongst us. I struggled out of the bloody mozzie netting, grabbing for my SA80 with the one hand and the TACSAT with the other. I slammed a round into the assault rifle's breech, and flicked the safety to the 'off' position. As I turned to face the enemy, I became aware of a circle of familiar faces at the arched doorway of our 'bedroom'. I couldn't hear a thing, for I'd been deafened, but I could see the lot of them pointing at yours truly and pissing themselves laughing.

For a second my confused and scrambled mind tried to grasp what was going on. Then I noticed that the underside of my camp bed was soaking wet, and that there were shards of plastic bottle scattered all around the floor. This was no enemy attack. I'd been MRE-bombed, and I didn't find it the slightest bit funny.

'Fuck off,' I mouthed at Sticky, Throp and the other lads. I was yelling, but I couldn't hear a thing. 'Fuck off and let me sleep, or I'll knack you.'

I clambered back inside my baking mozzie-tent and collapsed on to the camp bed. I tried to get my heartbeat back to something like normal. I cursed Sticky, for it had to be him. The MRE-bomb was a simple enough device, and one of my favourites. I reckoned it was class whenever I got the other lads, but it was never quite so funny this way around.

Our Meals Ready to Eat (MRE) Army rations came complete with charcoal sticks. If you wanted some hot nosh, you'd throw the stick into a pan of water, drop the bag in the pan, and the chemical reaction would get it cooking. But if you took an empty water bottle, stuffed the charcoal stick inside, added a little water and screwed the lid on, then you had a DIY bomb. You'd have to sneak it under someone's bed without waking them, and before the thing exploded. It'd go off like a rocket, and the victim would wake with a real flap on, just as I had done. Those bombs were truly, seriously loud, and nine times out of ten the victim would come piling out of his room with his weapon locked and loaded.

The best time to do it was just after stand-to, when the young 2 MERCIAN lads were trying to get a bit of extra kip. But this time it was me who'd been MRE-bombed, and I just couldn't get back to sleep. Eventually, I gave up trying and wandered over to get a brew. I did my best to ignore Sticky and Throp's smirking, and went to check in the radio room if anything was cooking with the enemy.

Busty came to join me in the shade of the 'radio shack', a canopy of camo-netting slung between the FST Vector and the main, mud-walled building of the base. We had a good natter about the previous day's fighting, and how the air war had worked alongside the ground war, and how winning that battle might have altered things in the Triangle.

The OC reckoned we'd hit the jackpot, taking out Commander Jamali. He figured we'd cut off the enemy's head, and left them running round like the proverbial headless chicken. Not a thing was happening anywhere in the Green Zone, and for sure something had knocked the fight out of the enemy. For days they'd been building up to this massive attack, which had clearly been aimed at taking Alpha Xray. Instead, we'd hit them again and again and again, and ended up killing their top commander.

I guess Jamali was their equivalent of the OC. I wondered for a moment how us lot would have reacted, had Butsy and his HQ element been taken out. It didn't bear thinking about.

The OC reckoned it would be fairly quiet until they got a replacement commander into the Triangle. Once they did, they'd start trying to probe our positions, as they had been doing before. They would try to establish the boundaries between our bases, and also any weaknesses they could exploit. And they would try to get behind our lines, so they could attack from all sides with us sandwiched in between. But for now, we should enjoy the quiet after Commander Jamali's killing.

Because for sure it wouldn't last.

EIGHTEEN
HELLFIRE'S THIRTEEN

It was stand-to. A couple of days had passed since the big battle, and the 2 MERCIAN lads were in their positions ready to rumble, should the enemy fancy having a go. Intel had been coming in thick and fast that the enemy were reinforcing their units, and getting into positions to attack. Same-old, same-old.

The sun rose harsh and brittle over the low mountains to the east. There wasn't a sniff of action from the Green Zone, so most of the lads went back to bed. I was chatting to Lance Corporal John Hill, one of the Somme Company TA blokes whose job was to manage the interpreters. John was a forty-year-old, barrel-chested publican who ran two boozers back in London. I had an instinctive liking for him. He and Jason Peach, the 2 MERCIAN sergeant major, were as thick as thieves, and John would take no shit off anyone, least of all the terps.

We were torturing each other with visions of a cold pint of ale, when Jase went quiet. He jerked his head in the direction of one of the HESCO perimeter walls.

'Here, lads, what d'you make of that fucker?'

John and I glanced where Jase had indicated, and there was that dodgy loner of a terp, the one we'd vowed to keep an eye on. As we watched, the guy paced out the length of the southern wall, counting out as he did so. Then we saw him starting the same with the eastern wall, murmuring numbers as he went.

We glanced at each other, and an instant later we were legging it across the compound. We dived on the terp, hurling him to the ground. As we held him down, Jase produced a length of paracord from somewhere, and we tied the guy up. We hauled him back to the Vector, and got Alan, our own terp, to have a few words with him.

'Tell him we're going to hand him over to his mates, the Taliban. Tell him that's who he must be working for, so he can go and join them.'

Alan did the translation, and the terp started protesting: 'No, no! I am innocent! I have been doing nothing.'

'What the hell were you up to pacing out the walls then?' we demanded.

The terp had no way of explaining his actions, so John decided the guy would have to be sent to FOB Price, for questioning. We plasticuffed him, securing his hands with plastic handcuffs. John put the guy under guard in the terps' room. He had his hands untied, but he was booked on the next helicopter ride out of there.

After dealing with the dodgy terp I needed a brew. I got one on, then Sticky, Mikey Wallis and I went up on the roof to enjoy the view. I lit up a tab, and inhaled deeply. Down below I could see Jase Peach pottering about in the compound. Jase was an excellent bloke, and a top soldier. Recently, one of those parcels had arrived from well-wishers in the UK, with a consignment of pump-action water pistols. Jase, Throp and I had taken to hanging out at the well in the heat of the afternoon. Whenever someone came over to have a wash or a cool-down, we'd ambush him.

Some of our victims didn't find it very funny – but we did, every time. They'd come to the well for a good cool-down, so what did it matter if it came from the well bucket, or our pump-action water guns?

I was just giving Jase a thumbs-up, ref the dodgy terp, when from out of nowhere there was a tell-tale violent burst of flame – a horizontal mortar flash aimed right at us. I didn't need to see the black streak rocketing across the valley, to know that we had an RPG inbound.

The three of us were down on our bellies in a flash and clinging on to the domed roof, not that it would help much – there wasn't a scrap of cover anywhere. Not one of us had our body armour or helmets on, and didn't we feel like bloody fools now.

The rocket-propelled grenade came drilling in, its pointy head looking like it was dead on to smash us. At the last moment it veered slightly upwards, howling past a few metres above our heads. I turned to see it ploughing onwards into the desert. It detonated with a punching blast out in the midst of the emptiness.

'Fucking hell,' I kicked off. 'I dropped me bloody tab.'

I had. It was only half finished, and I could see it smoking away on the edge of the roof. As I wriggled forward to grab it, there was the sharp buzz-snap of a high-velocity sniper round ricocheting off the roof. It had gone right between where Mikey and I had been.

Fuck me, was that close. We were off the roof like greased weasels and piling down the ammo-box steps, giggling our heads off. I still made sure I recovered my half-smoked tab though.

'Fucking snipers,' Mickey grouched. 'That bastard better not've punctured my radar dome.'

His weirdly shaped mortar-locating gizmo was stuck up on the roof right next to JTAC Central. Mickey stomped off to check on his computer that his radar was still working.

I got word that the lads in the front sangar had spotted the position from where the RPG had fired. It had come from a compound bang between us and Alpha Xray, and a little to the east. I got on the air and dialled up some CAS, to see if we couldn't catch that RPG team.

I got sent *Dude One One* and *Dude One Two*, a pair of F-15s. I had them flying search transects all around the position of the enemy RPG team, but nothing could be seen. They checked out of my ROZ at 0815, low on fuel.

I asked for more air, and kind of regretted it when I heard what I was getting. I had *Overlord Nine Seven*, a Predator, allocated to me for eight hours straight. The drone arrived in the overhead at 0900. I began a chat with the first American operator, knowing that I'd be staring into my Rover screen until 1700. I was ecstatic.

Seven hours later I was on my fiftieth brew courtesy of Sticky, and I'd smoked ninety tabs or more. My eyes were like a cow's udders, I'd filled up god only knows how many piss bottles, and not a sausage had we seen. It was 48°C in the back of the Vector and I was in my shorts and nothing else, but still I was sweating like a pig. Yet I didn't fancy going up to JTAC Central until we'd nailed that bastard RPG team.

I'd gone through three different controllers in Nevada or wherever they were, and I'd recced from Helmand to Whitby and back. I was not enjoying this. It was made all the worse by Throp and Sticky constantly ripping the piss. Each new American operator would start his shift with boyish enthusiasm, and I'd have to try to reciprocate.

'Visual one male pax walking down a dirt path between two trees,' the guy would announce, in his thick American drawl. 'Can you see him?'

In the background, Sticky was doing his best Yankee accent, mimicking the operator. 'Say, you know, y'reckon we got us Osama Bin Laden his goddamn self?'

I was trying not to laugh.

It was on the tip of my tongue to say, 'Yeah, that bloke on the path – he's an Afghan farmer going home to have his dinner.' Instead, I feigned interest and told the Predator operator to 'keep a close eye on him'.

Next guy alerted me to being 'visual two pax cooking their dinner'. I glanced at my watch: only thirty more minutes to go. At 1635 the operator announced he was visual with three pax walking down a track. I told him I was visual with them too. Then the three pax met up with three more pax, so now there were six.

I hunched a little forward in my seat. The Predator was at 22,000 feet, so there was no way the six figures would be able to hear or see it. The six pax met with seven more, and now there were thirteen. This was starting to get interesting. I told the operator to keep with the thirteen pax, then shouted out the door of the Vector.

'I've got thirteen pax two-seventeen metres east of Alpha Xray!'

At that point Chris and anyone else who could squeeze into the Vector gathered around. The thirteen guys on my screen were all dressed alike, in black turbans and black robes. Not a weapon could be seen, but that didn't mean a thing. The enemy were masters at hiding their guns until the very moment of attack.

'*Overlord Nine Seven, Widow Seven Nine*: thirteen pax together like this is highly unusual. What can your analysts make out of those figures? I'm specially interested if any of them is carrying anything resembling a weapon, even if it's hidden.'

'Roger. Stand by.'

Behind each Predator operator are a team of analysts who can rewind, pause, zoom in and flip the video images on a huge viewing screen. As opposed to our feed, they had the ability to scrutinise the footage in incredible detail.

'*Widow Seven Nine, Overlord Nine Seven*: my analysts say that every one of those thirteen pax is carrying a concealed weapon.'

What the fuck! I couldn't see even the hint of a gun. But if they were all armed, we had thirteen enemy fighters moving ever closer to the lads at Alpha Xray.

As I studied the grainy image on the screen, I saw one of the figures heft something under his robe higher on to his shoulder. It was a long, heavy weight, and it had the distinctive silhouette of an RPG. For an instant there was the glint of metal in sunlight from under the robe.

'Freeze that frame,' I yelled to the operator. 'Check out what that is shining.'

'Roger.' A pause, as the tape rewound. Then, 'It's an RPG launcher'.

'Aye, I reckon it is,' I told him. 'Stand by. Get the OC!' I rasped to whoever was nearest

I kept my eyes glued to the screen. The guy with the RPG started moving swiftly away from the main group towards Alpha Xray. Fuck. My instinct was screaming at me that they were massing to hit the base, and all I had on station was a lone Predator.

I got the operator to widen the field of view, so we could track the lone figure and the original twelve left behind. The solo fighter led us to another group of men. By now there had to be thirty or more massing beneath the trees some fifty metres short of Alpha Xray. I got on the air to Damo Martin, back in the FCP at FOB Price.

'Damo, I need immediate fucking CAS now,' I rasped. 'We're TIC-imminent and I'm visual with thirty-plus armed pax about to whack our lads at Alpha Xray.'

'Sorry, mate, there's nothing available,' Damo replied.

'You're fucking joking!'

'No mate, there's nothing. All available air is out on TICs.'

'Damo, you're not hearing me, mate: I really need some fucking air.'

'OK, I'll see what I can do. Stand by.'

I turned to Major Butt, who was standing in the doorway of the wagon. 'Sir, get on to the lads at AX and tell 'em to stand-to now. There's dozens of blokes with RPGs and shit fifty metres short of the base, coming in from the east along Route Buzzard.'

Butsy gave me the nod. 'I'm on it, Bommer.'

There was a squelch of static and Damo was back on the air. 'Bommer, mate, there's nothing. As soon as we have air, you'll get it. But we ain't got nothing now.'

'Fuck it, we'll use the Predator,' I muttered.

The trouble was we had scores of enemy to kill, and the Predator was carrying just the one Hellfire. Plus the males of fighting age were in two separate locations. I turned around to Sticky, Throp and Chris.

'Lads, here's the plan. Get the boys at AX to fire 51mm mortar rounds, but only smoke mind, three hundred metres beyond the junction of Routes Crow and Buzzard. The enemy will think they're under attack, and bunch together. At least, that's what I'm guessing. Before they realise it's only smoke, I'll smash 'em with a Hellfire.'

There was no time for discussion, as the Predator had only fifteen minutes' flight time left. As the lads got on the radio to the mortar team down at AX, I dialled up the drone's operator.

'*Overlord Nine Seven*, *Widow Seven Nine*: I want you to bank around to the south. But keep eyes on the enemy pax. Repeat: keep eyes on.'

As I spoke those words my Rover terminal started to flash, warning me that the battery was about to die. Of all the fucking times!

'Roger, banking around south,' the operator confirmed.

'OK, I want you to cue up for an attack run.'

There was a pause. 'Will you repeat that instruction, sir.'

'I want you to cue up for an attack run using your Hellfire.'

There was another, longer pause. 'Sir, we've never fired the Hellfire before.'

'I don't care. You're the only platform I've got and we've got friendlies about to be whacked. I need you to cue up for an attack run on those enemy fighters.'

'Sir, who is clearing me for this mission?'

'I fucking am – *Widow Seven Nine*,' I rasped. 'And we're using your fucking Hellfire.'

'Uh, sir, wait out.'

I could hear the chat in the background, as a bunch of American couch potatoes ran around excitedly telling each other they were being asked to actually fire a missile in anger.

'AX are ready to fire smoke,' Sticky yelled over to me.

'Fire on my order,' I replied. 'Chris, I need you to coordinate the mortars down at AX. I need them landing three hundred metres due east of AX on my call.'

'No problem,' Chris confirmed. 'Just tell me when you need them in the air.'

The bloody image on the Rover terminal kept crackling and fuzzing, as the Predator banked around. I kept thinking: we're going to lose the bloody downlink, or my Rover screens are going to pack up, or the bloody Yank operator's going to bottle out of doing the hit.

I mopped my brow with the back of my hand, and willed it all to come together.

'Overlord, this is Widow: are you ready, or what?'

'Sir, are you certain I am cleared to fire?'

'Too fucking right I am,' I confirmed. 'Chris – get the mortars in the air!'

'Roger!' I heard Chris give the order: 'Engage with mortars on direction and target given.'

A second or two later, the image on my Rover screen showed the bloom of an explosion two hundred metres beyond the enemy positions, just as I'd asked. As the mortar rounds hit and started gushing, a cloud of grey-brown smoke drifted lazily across the terrain.

I saw figures running. The fighters were bunching together at the centre of their mass. More and more streamed in to that one position. The plan was fucking working! They thought they were under attack, and were grouping together to muster their response. But it wouldn't take long for the enemy to realise it was only smoke rounds.

'*Overlord Nine Seven*, you're clear to fire on the concentration of fighters beneath those two trees.'

'Sir, I need thirty more seconds,' the operator replied.

'You what?' I practically screamed.

Ten seconds of silence followed, each of which dragged like a lifetime. At any moment I was expecting the fighters to disperse, and to launch their attack on AX.

'Sir, I'm ready now. Can you confirm I'm cleared to fire the Hellfire?'

'Fucking right you are,' I yelled. 'Just fire the bastard thing! You're clear hot. Engage! Engage!'

An instant later the image on the Rover terminal collapsed into a pixillated mess. I guess the Predator had fired, and the kickback of the Hellfire had given the drone a massive speed-wobble. Either way, the image had gone to rat shit.

'Have you fired?' I yelled.

'Affirmative,' the operator replied, with something like real satisfaction. 'One Hellfire missile is on its way, sir.'

The image came back again. As it stabilised I felt my heart miss a beat. The enemy figures had disappeared.

'Where the fuck've they all gone?' I yelled into my TACSAT.

'Sir, stand by.' There was a pause of a few seconds. Then: 'Sir, the enemy pax are underneath the two trees. My analysts can see at least twelve pax under the foliage.'

With the Predator at 22,000 feet, I reckoned the Hellfire would take a full thirty seconds to reach target. All eyes were glued to the screen. Not a word was spoken, and I was holding my breath as the missile streaked in.

For an instant there was the lightning flash of a black splinter streaking vertically through the screen, and then it hit. It was smack bang on target, the Hellfire ploughing into the earth right between the two trees.

'Splash! BDA from analysts,' the operator's voice came up on the air. 'Five have been killed outright. Many injured.'

'Roger that, Overlord. Good strike.'

'Sir, I have to leave your airspace. Low fuel. I have to leave.'

Eight hours was the maximum air-time for a Predator, and I'd had him for seven-fifty as it was. I guessed the operator was shitting himself that his multi-million-dollar aircraft was going to run out of gas and ditch in the Afghan wildlands.

'Roger that, Overlord. But keep your pod on the target area as you fly out of my ROZ.'

I wanted eyes on for as long as possible. As I gazed at the image, four figures scurried into view. They laid a blanket on the dirt, and started piling something on to it. As I watched, I realised it was blown-up pieces of human being. It was body parts. It was arms and legs and chunks of human flesh.

It wasn't very pleasant, but it was either them or us, and this time the fight had gone our way. Three figures came crawling out from the trees. One of them lurched forwards on to the path, then lay completely still. I guess another was dead.

I sent a sitrep to Damo Martin: '*Widow Eight Two*, *Widow Seven Nine*; one times Hellfire fired and five times KIA Taliban ...'

'Yeah, yeah,' Damo cut in. 'You're trying to tell me you've killed five enemy with one Hellfire.'

'Listen, I used the Predator's Hellfire and I've got the footage to prove it.'

I had. It was another great thing about the Rover terminal: it recorded the images of the strike.

Damo came back, 'Bommer, if what you're telling me is true, it's bloody class, mate: I want a copy of that bastard footage by teatime.'

The feed from the Predator was breaking up now.

'Overlord leaving your ROZ,' came the pilot's voice. 'Thanks very much, sir.'

'Aye, it was fucking awesome mate.'

'Yes sir, that it was. It was good working with you, sir. You stay safe down there, *Widow Seven Nine*.'

'Aye, I fully intend to, mate.'

I flipped frequency to Damo Martin's. 'ROZ cold,' I told him, signalling that there was nothing more happening in my airspace right now.

The chatter was going wild, as the enemy called for units to check in, but there were few if any answers. The OC was chuffed as nuts with that Hellfire strike. It was the first time the Green Army had ever fired a Hellfire from a UAV. To have done so, and taken out so many of the enemy, was one hell of a strike. The analysts in the US reckoned we'd killed seven outright, six were fatally wounded, and there were a lot more injuries.

At a minimum it was thirteen to that one Hellfire, which wasn't bad going. We reran the footage on the Rover screen, and more details became clear. The enemy had outlying sentries posted, some of whom were less than fifty metres from Alpha Xray. Even when they'd mustered under the mortar fire, those sentries had remained in their positions.

It looked as if that one Hellfire strike had scuppered a big push on Alpha Xray. At 2000 hours I got a pair of Harriers overhead, and for two hours I had them flying search transects over the Green Zone, but it was deserted. I hit the sack and slept like a dead one until stand-to.

NINETEEN
GET SNOOPY

Chris kept twisting on about Sticky's Snoopy key ring. The day after the Hellfire strike, he stuck his head into the back of the wagon. Sticky's pack was on the seat, and there was the Snoopy dog hanging off the back of it. Chris stared at it for a second or two, then gave Throp and me the look.

'Lads, I want something to happen to Snoopy,' he announced. 'Get fucking rid of it.'

Throp and I looked at each other. Fair enough. Chris was the boss, after all.

'Leave it with us, mate,' I assured him.

'Consider it done, boss,' Throp added.

We shut the Vector's door, and I fished out a big black marker pen from my JTAC kit. The two of us proceeded to draw big black knobs all over Snoopy's white fur. Then Throp unhooked it from the pack and dunked it in my urine bottle. That was it – job done.

Throp hung it back on Sticky's pack but the wrong way round. You couldn't see Snoopy's face any more: all you could see was the dog's back covered in knobs. A little later Sticky was back with us and fiddling around with his kit. All of a sudden he noticed that his Snoopy dog had been horribly defaced. He threw a track.

'Who the fuck's been drawing knobs on my Snoopy?' he snarled.

I turned away from him trying to hold the laughter in. I grabbed my mug of lukewarm tea, took a slurp and got my eyes on my Rover screen. Throp just stared at Sticky with that look of his: *What am I accused of doing this fucking time?*

'It's fucking Chris, isn't it?' Sticky raged. 'Ever since I got it he's hated that Snoopy.'

Throp shrugged. 'Fuck knows. All I know is it wasn't us.'

Sticky reached out for the toy dog, took it off his pack and put it away carefully in his pocket. He had a right arse on. I'd never seen him so angry.

He glanced at Throp and me. 'Look, lads, all I want to know is who drew cocks on my dog?'

That was it. I cracked up laughing, spluttering tea all over the wagon. All that morning Sticky kept asking the same question – *Who drew knobs on my Snoopy?* As he wasn't getting any answers, he just concluded that it had to be Chris. It was fair enough, really. After all, it was Chris who'd ordered me and Throp to get rid of it.

But it was Jess, not Sticky, who was becoming the real victim of the FST wind-ups. Jess was the newcomer on the team, so I guess it was only natural for him to get picked on. He was a good lad, but he used to bite easily. And there was this weird clash between him and Chris that we reckoned all boiled down to hockey.

Jess had one glaringly obvious shortcoming: he didn't seem able to grow a proper beard. By now the four of us had manly beard-fungus sprouting all over our faces. We hadn't shaved since arriving at Monkey One Echo, two weeks earlier. We all had proper monster beards coming on. But for some reason Jess only seemed able to manage a light dusting of fuzz under his chin, with a couple of sprouts to either side of his mouth. Chris had nicknamed him 'Upside Down Beard', and Jess bloody hated it.

Sticky and I had got into the habit of tying up Jess's clothes, whenever he was away at the well. We'd get his trousers, knot them, then each grab a leg and run in opposite directions. Then we'd do the same with his shirt. Jess would come back from the well and try to get his trousers on, only to find a knot in them the size of a boiled egg.

We'd be sat in the back of the wagon, pretending we were seriously busy. Jess would come over twisting. The four of us would look away, trying to pretend it was nowt to do with us, and doing our best not to crack up. How he managed to get those knots undone I will never know. Plus we were always hitting Jess with MRE-bombs.

With that lone Hellfire strike having smashed the enemy, there was nothing doing from their side, so we decided to have a big cricket day. We each kept the score in our heads and no one tried to cheat. If someone kicked off saying the ball hadn't hit the stumps, and they had twenty of us saying they were out, then they were out. It was a great way to keep fit and to have a laugh, and to get a good suntan.

At 2200 I got allocated air, having two F-15s for three hours on yo-yo from a refuelling tanker. I got them scanning the length and breadth of the Green Zone, but there was zero happening. There wasn't the slightest hint of an enemy presence anywhere. If you weren't careful you could forget there was a war on. The total lack of enemy activity was spooky. That one Hellfire strike from the Predator couldn't have put them all out of action. So where were they? And what were they up to?

The following morning we had two journalists pitch up at PB Sandford. They'd come down on a road move with a big resupply convoy. John Bingham was the writer, and Andy Parsons the photographer. Andy was the younger of the two, around twenty-seven or twenty-eight, and he was a top lad. He turned round to us on that first morning, and said: 'I'm here to take photos, and John's here to write stories. So watch what you're doing and watch what you're saying around us.'

It was fair enough, and I appreciated the up-front honesty of the bloke. They carried with them blue body armour and helmets, to mark them out as being press. I didn't think it would make the enemy any the less inclined to kill or capture them. But they had their rules, just like we had ours, and one of theirs was that they wore blue kit when going to war.

At last light I was up on JTAC Central with a pair of F-15s in the overhead. I was determined to find some enemy, if only to show the reporters we really were at war. From up on the roof I could sense they were out there, watching and waiting, ready to strike.

I'd just tasked the Dude call signs to fly search transects the length and breadth of the Green Zone, when they got ripped away to a TIC up at Kajaki, in the far north of Helmand. There was no more air available, so I went to bed none the wiser about where the enemy might be, or what they were planning.

After stand-to the following morning Butsy gave the order to take a foot patrol down to Alpha Xray, and Sticky and me were on it. The main purpose was to show a presence in the Triangle, and to check on the lads down at the Alamo.

Over the past couple of days there'd been a bit of a nasty fright on at AX, with bouts of horrible sickness. The OC had pushed a convoy of WMIKs and Snatches down to the base, and extracted the sickest of the lads, who was airlifted to Camp Bastion. A visit from a foot patrol would cheer their morale. Plus we could give the two journos a look-see around the Triangle.

We set off at 0900, taking Route Buzzard east along the high ground to Monkey One Echo. As soon as we were out of the gates, the radio chatter was buzzing that the enemy were visual with us. From Monkey One Echo we hooked south towards Route Crow, and were sucked into the suffocating humidity of the Green Zone. As we pushed ahead, well spaced in single file, I could feel eyes in the bush to either side of us.

We made Alpha Xray without incident, and filed into the base. I'd been down at AX quite a bit now, and I supposed I'd got used to it. But the reporter's eyes were out on stalks. If PB Sandford was luxury, then Alpha Xray was the fucking pits. It was a place for fighting, eating, sleeping and defecating, and that was about it. It was the Alamo transported to the wilds of Afghanistan in the midst of a twenty-first-century war.

The entry point into Alpha Xray was a mud bridge over a shallow canal running along Route Crow. The bridge terminated at the main building, a two-storey square structure. Its thick mud walls were pitted with bullet holes and RPG craters, like a bloody great big sieve. Strung around the base of that building were rolls of razor wire, and the one gateway in was barred with coils and coils of the stuff. Inside, there was a rectangular compound, hemmed in by thick mud-brick walls twice the height of your average bloke.

The sandy-floored compound was doss house central for the twenty-odd lads garrisoning AX at any one time. It was threaded across with makeshift washing lines, slung with socks and pants hung out to dry, with heaps of kit stuffed into the thin shadows at the base of the walls. In one corner of the compound a blue Fosters lager sunshade had been erected over a rickety garden table. To either side the rectangular windows in the walls had been filled with sandbags, leaving just a slit of a gun turret through which to put down fire. There was a palpable sense of the siege about the place, and after their four-day stint none of the platoons was loath to leave.

A rank of body armour, backpacks and helmets lined the wall nearest the main building. Leaning carefully against each set of kit was an SA80, GPMG or Minimi Squad Assault Weapon (a drum-fed light machine gun). Belts of ammunition were wrapped carefully around the weapons, to keep the links out of the dirt. A rickety wooden ladder led up to the flat rooftop above. It was from there that the real defending, and the killing, was done.

The rooftop position at Alpha Xray had all-around vision over the surrounding bush. The three sides of the position looking away from the compound were lined with sandbag walls, to give a prone or crouching soldier some cover. A battlement in each corner provided a fire-turret for the GPMGs, and the lone 50-cal on its chunky tripod mount. Comms antennae bristled from radio packs on just about every corner. Piles of spent shell casings littered the roof, testimony to the ferocity of the recent fighting.

Scattered in amongst them were empty plastic water bottles and discarded helmets and crates and crates and crates of ammo. From the rooftop, the nearest treelines and thick cover were spitting-distance close. You could get around eight soldiers up on the roof, and that was the backbone of the defences here at Alpha Xray.

We had a chat with the lads, and all seemed to be bearing up well. Then we set off on the return leg of the patrol. Sticky and I were halfway across the river bridge on Route Crow, when Alan the terp alerted us to an item of radio chatter.

'They're saying they can see three figures on the bridge,' Alan muttered. 'They're saying the central one has the stubby black aerial that controls the aeroplanes.'

Oh fuck. I hurried across the bridge with a feeling like ice running down my spine. I felt like someone had a sniper's crosshairs bang on my head. It was horrible. But not a shot was fired. We reached PB Sandford without having been engaged, and herded through the gates. I just couldn't understand why they hadn't whacked us.

The OC couldn't understand it, either. He'd pushed a big looping patrol all through the Green Zone, from the eastern side of Rahim Kalay to the western edge of Adin Zai, yet not a sniff of the enemy. Not a shot had been fired since that Predator's Hellfire strike.

Where the fuck was the enemy? What were they up to? What were they planning?

The OC gave orders to push out a second foot patrol, for the early hours of the following day. This time, we'd head out from PB Sandford as two full platoons, and at night. And we'd press further eastwards than ever we had before, into real bandit country.

The eastern limit of the patrol was to be an enemy position marked as Golf Bravo Nine Eight on the GeoCell maps. It was a full half-kilometre beyond Alpha Xray, and totally uncharted territory as far as we were concerned. The patrol would consist of both platoons from PB Sandford, so forty-odd men, plus the OC and his HQ element and the full FST. We'd leave only a skeleton crew behind, and we'd link up with more lads at Alpha Xray. Plus we'd take the two journalists, Andy and John, with us. If that didn't get a rise out of the enemy, then nothing would.

At 0130 I had my first air checking into ROZ Suzy. I had two A-10s, *Hog Zero Seven* and *Hog Zero Eight*. I tasked both aircraft to fly air recces over the vanguard of the patrol, as we pushed south and east into the Green Zone.

At 0200 we filed out of PB Sandford in the pitch dark, a long snake of heavily armed fighters on foot, and all on night-vision. Apart from the clink of gunmetal on body armour, and the jet-whine of the A-10s high above, the valley was utterly still and silent. But we knew the enemy were out there somewhere, and we were going hunting.

By 0320 we'd advanced a good four hundred metres past Alpha Xray, and were deep into unknown territory. We were moving ahead at a dead slow. A dozen paces, then the whispered order to halt passed along the line. The entire column would remain motionless for a minute or more, crouching and listening intently in the hollow, ringing silence.

We pushed ahead for a good twenty minutes or so, before a cry rang out in the darkness. It came from the direction of our front, and it sounded like a verbal challenge in Arabic. An instant later, the night exploded all around us, as a barrage of RPG rounds came howling out of the trees and slamming into the bush. I dived for cover, brought up my SA80 and was about to open fire, so strong was the soldier's instinct to get the rounds down and to fight. Instead, I forced myself to grab Sticky by the arm, as I yelled into his ear.

'Get the point platoon's fucking coordinates!'

As Sticky got on his radio, I was hunkered down against a tree trunk and bawling into the TACSAT. Rounds were fizzing through the air above, and all around me our lads were putting down a savage amount of return fire. It was deafening.

'Hog call signs, *Widow Seven Nine*. Sitrep: under massive contact RPGs and small arms. Stand by to attack.'

'Roger. Standing by.'

Sticky had his face in mine, yelling out the grids.

'*Hog Zero Eight*, enemy forces are eighty-five metres to the front of our lead platoon, in a treeline running between Golf Bravo Nine Three and Golf Bravo Nine Five. Treeline runs for two-fifty metres, in a south-east to north-west dogleg. Friendly grid is: 93850269. Readback.'

The pilot confirmed the grid.

'I need a 30mm strafe on enemy treeline, on south-east to north-west attack run, to keep it away from friendlies. Repeat: friendlies are eighty-five metres to the south-west of target, danger-close at night.'

'Affirmative.' A pause. '*Widow Seven Nine*, I'm visual with the leading edge of your platoon. Plus I'm visual on the IR with three heat sources in the enemy position, plus ...'

The last words of his message were lost as an RPG round tore through the bush above me, exploding with a massive boom. It sounded horribly close, as the walls of vegetation threw back the raw crunch and slam of battle clatter in a deafening wave of sound.

'I repeat,' came the pilot's voice, 'now visual five heat sources in enemy position. Tipping in.'

'Roger. Nearest friendlies eighty-five metres!' I yelled. 'Nearest friendlies eighty-five metres west of target!'

Above the battle noise I caught a few scrambled words of Chris putting out an all-stations warning about the strafe, and then the pilot was radioing for clearance. I couldn't see him through the crush of vegetation, and I sure as hell couldn't hear him. I'd have to clear him in blind.

'No change friendlies!' I yelled. 'Not visual your attack! Not visual! Clear hot! Ground Commander's initials SB.'

'In hot.' A beat. 'Engaging.'

Above the staccato roar of the battle, there was a new noise now – the howl of the diving jet, and the purr of its Gatling gun pumping out the 30mm shells right above us. The instant it had finished firing Sticky had the lead platoon commander on the air.

They could hear screams coming from the positions to their front, where the A-10 had hit. That had to mean there were enemy wounded.

'*Hog Zero Eight*, good strafe,' I yelled. 'We hear screaming from that enemy treeline – their wounded. I want immediate re-attack, same position, same line of attack, no change friendlies, and danger-close.'

'Roger. Banking round.' A pause. 'Tipping in.'

The A-10 put in a second, much longer strafe, for several seconds the awesome throb of the seven-barrelled cannon drowning out the battle noise. His confidence had been boosted by the first strike being smack-bang on target. This time he had a BDA for me: two pax confirmed dead. Fucking great news: we were starting to win the firefight.

The patrol had gone firm in its positions, awaiting further orders from the OC. I got the A-10 pilot searching in the treelines to the north of where he'd strafed, scanning with his IR scope for heat sources. Within seconds he came back to me with this.

'Visual six pax north of the treeline targeted, and to the east of an L-shaped compound. Visual muzzle flashes from out of that position.'

'I want immediate attack with 30mm,' I told him. 'I want both Hog call signs shooter-shooter, on a north to south run.'

'Shooter-shooter' meant the aircraft would be coming in sixty seconds apart, with the first A-10 strafing, and the second doing a follow-up strafe as soon as the first was off target.

'Negative, *Widow Seven Niner*. I need *Hog Zero Seven* to keep a watch on my wing. I'll do two runs at the same time: one at altitude, and one as I'm closer in.'

This was fast and furious now. I was asking the pilots to do repeated danger-close strafes at night – the most challenging and risk-laden airstrikes possible. The pilot was going in more or less blind, over a dark and confused battlefield. He needed his fellow pilot – his wing – to watch over him, guiding him in as he did two strafes from altitude.

'Happy with that,' I confirmed. 'Friendlies eighty-five metres danger-close.'

I cleared him in. Chris put out the warning of a double-strafe, and for all the lads to get their bloody heads down. The A-10's dive was a long one, and the first 30mm strafe rumbled through the dark night like a distant, rolling peal of thunder.

A few seconds later came the second, the long throaty roar of the cannon closer and more threatening. But as the Gatling gun ceased firing, I could hear the coughing of the A-10's big jet engines, set high and ugly on the aircraft's tail. I knew in an instant what had happened. The kick back from the two strafes had caused the A-10's engines to stall. As the aircraft plummeted earthwards the pilot was having to try to restart his engines in mid-air. For a second or more I held my breath, and then the reassuring jet-whine cut through the night again. The aircraft picked up power, howled out of its dive and thundered low and fast across the valley.

Phew. Thank fuck for that. As the scream of the A-10 died away, I realised that all around us the bush had fallen silent. After those two mega-strafes, the firefight had died away to nothing. I breathed easily for a moment, then dialled up the A-10 pilot.

'*Hog Zero Eight*, awesome strafe. Requesting BDA.'

'BDA: lots of tiny heat sources in the treeline, but no further movement. Six pax dead. And *Widow Seven Niner*, BDA almost included one US pilot. I pretty much lost my engines for a second there.'

'I know, mate. I heard it. Thanks. It was class. It was the best strafe of the tour.'

Alan was on to us now about the radio chatter. The enemy were going bananas. There were repeated calls for various units to check in, but no answers. After those four monster strafing runs from *Hog Zero Eight*, I wasn't particularly surprised.

I glanced upwards. The sky to the east was lightening with the first rays of dawn. I caught a glimpse of the stubby silhouette of an A-10 circling above. I said a quick 'thank you' to the pilots. They'd smashed the enemy in a danger-close battle at night, and in the midst of a murderous ambush on our patrol. It was a top job, to put it mildly.

Butsy gave the order to move out. We were pushing onwards towards our objective, four hundred metres further into enemy territory.

TWENTY
AMBUSHED, SURROUNDED, TRAPPED

We moved off on foot into the bush. It was 0500, and all around us the terrain was lightening, as the sun clawed its way over the hidden horizon. We'd lost the cover of darkness.

The A-10s got ripped by a pair of F-15s, call signs *Dude One Three* and *Dude One Four*. Alan was warning us that the airwaves were going wild. Enemy commanders were yelling at their men that we were 'coming in on foot'.

'Hold your positions!' they were ordering. 'Hold your positions! Do not attack yet!'

For forty-five minutes we pushed onwards into alien territory, in tense silence and alert to the slightest movement around us. One patch of bush looked pretty much like any other, but we were acutely aware that none of us had been this far east before.

Butsy called a halt at a deserted compound, so we could take a breather and orientate ourselves. I grabbed my map and checked my GPS. We'd pushed east as far as Golf Bravo Nine Five, and our final objective was no more than two hundred metres further on. I was dying for a smoke. I stuck a fag between my lips and sparked up.

A few metres away from me one of the 2 MERCIAN lads was busy re-bombing his mag. His fingers fumbled and he dropped a bullet. He bent to pick it up, and as he did so this flaming projectile came roaring through the window where he'd just been standing. It screamed over his back, tore across the space in front of me and slammed into the back wall. The RPG warhead buried itself in the mud-brick structure, exploded and smashed the compound wall to smithereens. As the choking cloud of smoke and dust cleared, the 2 MERCIAN lad was left sitting by the window, completely unharmed. As for me, my fag was in the dirt at my feet still smouldering away, but other than that I was perfectly all right.

I picked it up with shaking hand and clamped it between my teeth.

'Fookin' hell,' I muttered. 'Dropped me tab.'

The 2 MERCIAN lad shook his head and gestured at his ears. He was totally bloody deafened. But if he hadn't stooped to pick up that bullet, the RPG round would have torn his head off and exploded right in front of me. He knew it. I knew it. And the two of us were left staring at each other with eyes like bloody saucers.

We moved out and pressed onwards into the bush. A few minutes later a long burst of gunfire tore apart the tense silence. It was the rattle of an AK-47, and it was answered an instant later by an SA80. Suddenly, there were RPG rounds smashing into the bush all around us, as all hell broke loose.

I dived for the cover of a ditch. Sticky, Throp, Chris and Jess landed next to me, as we tried to work out where the enemy were firing from. Then, an all-stations call went out on the company net:

'Man down! *Arsenic Two Zero*, man down!'

Arsenic Two Zero was the call sign of 2 Platoon, on point. We'd been ambushed at close quarters and the boys up front were getting smashed. I felt that horrible, sickening feeling of knowing we had a man lying out there somewhere in the bush injured or dying.

'Get 2 Platoon's fucking grid!' I yelled at Sticky. I was having to scream to make myself heard over the battle noise.

'*Dude One Three, Widow Seven Nine,*' I radioed the F-15. 'Sitrep: under assault. Contact is raging hot, and we have a man down. I need you visual with the lead elements of our patrol, so you can find the enemy fire positions. Stand by for grid.'

'Roger. Standing by.'

I scrabbled around in the pocket of my combats and pulled out my battered map. It was 'Fabloned' – coated in a plastic film – but still it had taken a real beating. I spread it out in the shadowed damp of the ditch. I tried to block out the battle noise and the red mist of anger, as I searched for our location.

There would be time for rage and fury later. We had a man down, and we had to get him out. I glanced at my wrist GPS, and traced the coordinates on the map. We'd pushed so far east we'd fallen off the right-hand edge of my regular map – OP AREA 1 Ed2. I grabbed a second map, and found us. We'd gone beyond Golf Bravo Nine Six, with Golf Bravo Nine Seven to our south. Our objective, Golf Bravo Nine Eight, was a hundred metres to our front.

Sticky cupped his hands and yelled the grid of the lead platoon in my ear hole. '5-9-3-6-8-2-1-9.'

'*Dude One Three, Widow Seven Nine,*' I screamed into my TACSAT. 'Patrol is strung out between Golf Bravo Nine Six and Golf Bravo Nine Eight. Most forward grid is: 59368219. Readback.'

The pilot confirmed the grid. He was having to yell to make himself heard too. 'Visual muzzle flashes all around your lead platoon,' he reported. 'Enemy has your lead friendlies surrounded danger-close on three sides. They're maybe ten, twenty metres away from your guys.'

'Roger. Stand by.'

What the fuck! My mind was racing. We had our point platoon surrounded to the north, south and east, danger-close. Fuck danger-close – it was ten metres away. There was no way on earth that I could use the air.

'OC! Chris!' I yelled. '2 Platoon is surrounded ten metres on all sides but our own. I can't use the fucking air!'

For a second the three of us stared at each other, as the full implications of what I'd said sunk in. Then the OC was on the radio.

'*Charlie Charlie One*, all stations. Orders: full platoon assault to relieve 2 Platoon and extract casualty. Fix bayonets. Advance on my order.'

It was the only decision to have made. We'd do a fighting advance to reach 2 Platoon, and get the casualty out that way. All around me there was the sound of steel blades rasping on steel barrels, as the lads slotted their bayonets on to their weapons.

Nothing could ever bring home how desperate the fight had become more than the order to 'fix bayonets'. When it came to hand-to-hand fighting at close quarters, the air was of no use at all. I gathered up my maps and shit and stuffed them into my pockets, then rammed the razor-sharp dagger of my own bayonet on to the barrel of my SA80.

'*Charlie Charlie One*, all stations: platoon assault go!'

Butsy gave the order and we surged out of the ditch. Chris took point as we pounded ahead in an adrenaline-fuelled charge, kicking through the dust and rocks ahead of us. As we surged, the section to our front put down a savage wall of fire on to the bush to either side.

We charged ahead for fifteen metres, went firm, and started blasting away, as we gave cover for the section behind to come rushing forward. Up ahead the track hit a dense wall of trees strung with vines and thorns, and there it died. It was fucking carnage.

We piled into a rat run, a stinking, shallow ditch full of God only knows what. We crawled along it on hands and knees, as the rounds tore across above. We hit a flooded section and we were up to our waists in thick, foul-smelling black water and shit. We struggled ahead, staggering over submerged boulders and rotten posts in the

Plan of attack. Chris and the lads mapping out a
mission at a pre-op briefing, Patrol Base Sandford.

In the shit. Men of B Company 2 MERCIAN patrol
through a stinking ditch in the heavily vegetated Green
Zone just hours prior to the ferocious ambush that
became known as the Saving Private Graham incident.

Hours into the ferocious firefight east of Alpha Xray, exhausted 2 MERCIAN soldiers haul Private Davey Graham – shot four times in the stomach – out of the bush and into an open field in the heart of the Green Zone, so the Chinook can come in and airlift him to Camp Bastion.

Fast air. The warplanes that I called in to smash the enemy: an F16 on 'yo-yo' refuelling mid-air, prior to returning to battle

A10 Warthog ground attack aircraft, my platform of choice if Apache gunships weren't on hand.

An F15 taxis prior to takeoff: *Dude 1/3* and *Dude 1/4*, two American F15 pilots, would save the lives of dozens of us lot in the heart of battle.

An unmanned Predator drone – fantastic for getting eyes on the Taliban. With the help of Predator *Overlord 9/7* I killed 13 Taliban with one Hellfire missile.

The Light Dragoons B Squadron operating in north Helmand.
This is much like the resupply convoy that I travelled with deep into the
Helmand deserts. Dubbed 'operation mine strike' the convoy duly hit
several mines, as I got the air cover to hunt for the mine-laying teams.

FV107 Scimitar and Spartan armoured vehicles of The Light
Dragoons provided fast moving and fearsome firepower on
a number of operations throughout Helmand province.
I qualified as a Scimitar commander and fought in one during
the Iraq campaign before training as a JTAC.

'Woofer'. After a night in a howling sandstorm, a stray dog turned up at our Vector armoured vehicle, whilst it was parked up at Monkey One Echo. Flea-ridden and half starved to death, we soon had Woofer glowing with good health. British soldiers appreciate loyalty, and Woofer stayed with us during the entire time we were under ferocious siege in The Triangle.

And they're off! Throp adopted a local donkey and used to thrash it around the desert outside PB Sandford as if riding the Grand National.

The moment of detonation – a GBU-38 500-pound JDAM (Joint Direct Attack Munition) smashes into an enemy target at night, around Alpha Xray.

An FGM-148 Javelin man-portable 'fire-and-forget' missile, with graffiti: a devastating weapon when in the hands of our missile operators, during the ferocious siege of the Triangle.

Taking the shot. Up on JTAC Central posing for a photo with a pair of Apache gunships behind us, the enemy decided to have yet another go and started hammering rounds around our ears. We were off the roof like greased weasels. From left to right: Sticky, Chris, me and Throp.

© Andrew Parsons

Getting a brew at PB Sandford. Left to right: Chris, Sticky, me and the OC, Major Stewart Hill. Some bastard had pinched my mug, so the new one had a warning taped to it: FAT JTAC'S MUG: TOUCH THIS AND I'LL JDAM YOU.

All for a good cause. The Czech Army blokes were monsters, so when they decided to get their tackle out and get naked for the day, who were we to argue with them?

Mikey Wallis, the mortar locating radar dude, with the remains of a monster 120mm mortar, scores of which the enemy lobbed into our base. One went off 20 yards away from Sticky and me and by rights the two of us should have been shredded.

The men of B Company 2 Mercian, just prior to being relieved at PB Sandford by the Danish battle group. To the left of the 2 MERCIAN flag stands the OC, Stewart Hill, and behind the flag right in the middle are the five of us – me, Jess, Throp, Chris and Sticky.

shadowed half-light. For fifty metres we fought our way forward, each step taking us closer to 2 Platoon and our casualty. And then we stumbled into a solid wall of fire.

Rounds shredded leaves and branches all around us, and RPGs exploded on top of our position. In an instant I hit the deck, but I hadn't done so voluntarily. I'd been slammed down like Mike Tyson had thrown his biggest ever punch at me. The impact had smashed me in the top of my back, hurling me on to my face.

I came to my knees spitting out mud and dirt. It felt like a bloody great big mule had kicked me in the shoulder. I couldn't figure out what the hell had happened. I wasn't dead and no limbs were missing, or not that I could feel. I groped around the top of my body armour at the back, but I didn't seem to be pissing out blood from anywhere.

I shook the confusion out of my head, raised my rifle and started cracking off rounds. No time to worry about it. We were in the fight of our lives. I'd landed in a shallow ravine, and a storm of bullets was slamming over the top of us. All around me the lads were hunkered down in cover, and trying to return fire.

There was a cry on the radio net: 'Man down! Man down! *Arsenic Two Zero*, two more injured! But still fighting!'

Oh shit! We were three men down now. We'd stumbled into the mother of all ambushes, and we were getting smashed. I felt a desperate, insistent tugging on my left arm, the one that was cradling the front-grip of my SA80. It was Alan, the terp, and he was yelling something at me. I guessed it was a vital bit of Intel.

'Bommer, we have to get out of here!' Alan screamed. 'The Taliban – they are everywhere! All around us! They will …'

'Shut the fuck up!' I yelled back. 'Get a bloody grip, Alan. Get a grip!'

His eyes were wide with fear. I didn't blame the poor sod. He was a civvie, not a soldier, and this is what we had led him into.

All of a sudden it went deathly quiet. One moment the enemy had been blatting away, the next they'd ceased firing. With nothing to aim at, our lads stopped shooting. The blue-grey smoke of RPG rounds and of burned cordite hung thickly in the air. We glanced at each other, wondering what the fuck was happening now.

A cry rang out from the bush just to the north. There was an answering cry from the south. These were enemy voices, and they were moving past on either side of us. They would know every ditch and treeline here. As we blundered about like the proverbial bull in a china shop, so their fighters were slipping by unseen.

This was the moment when they went to outflank and surround us. At the same time they were getting beyond danger-close with the patrol, so we couldn't use the air.

'Dude call signs, *Widow Seven Nine*: we have enemy all around us.' For some reason I was whispering. 'Tell me what you see.'

'Stand by,' the pilots replied.

Then: '*Dude One Three*: I'm visual with at least a dozen enemy moving in from the north-east of your position. More coming from compounds to your north – all armed.'

'*Dude One Four*: I'm visual with large numbers of pax moving in through the treelines. Your forward platoon is totally surrounded. They're fucked.'

I felt like saying *Thanks for that, you dumb Yank twat*. But instead, I passed the F-15 pilots the instructions that I never thought I'd hear myself saying as a JTAC.

'Dude call signs, *Widow Seven Nine*,' I rasped. 'We're in a Broken Arrow situation. Repeat: I'm calling a Broken Arrow.'

'Affirm: you are Broken Arrow,' the pilot replied.

'Broken Arrow.'

By declaring a Broken Arrow, I'd torn up the rulebook. Broken Arrow means friendly forces about to be overrun and killed or captured. It clears the fast jets to do whatever the JTAC asks, even if that includes dropping ordnance on their own men to prevent them getting captured.

'I've called a Broken Arrow!' I yelled to the OC. 'I'm bringing in a danger-close strafe – like right on fucking top of us.'

The OC gave me the nod. 'Get 'em out, Bommer. Whatever it takes, just get 'em out.'

We were deep in the shit and getting deeper by the second. We were hundreds of metres into the Green Zone, cut off from friendlies and surrounded. We had three injured lads, and we had no idea how serious they were. The OC had given me the word: whatever it took, we had to get our wounded men out of there.

'*Dude One Three*: this is the grid of our most forward platoon: 59368219. Repeat: 59368219. That is the friendly grid. Readback.'

The pilot confirmed the grid.

'Attack instructions. I want a 20mm strafe to the north of grid, and I want it twenty-five metres from the friendlies. Attack run west to east. Confirm.'

The pilot confirmed the attack instructions. Calling for a 20mm strafe at twenty-five metres from friendlies was pretty much bringing it in on top of our lead platoon. But I didn't see that I had any options.

'Banking around now,' the pilot radioed.

'Roger.'

The OC mouthed into his radio headset. '*Charlie Charlie One*: all stations keep low. Jets coming in on strafing run.'

I heard Chris repeat the warning. At the same time the OC, Chris and Jase Peach were frantically working the radios, trying to get the IRT launched and a Chinook in the air to evacuate our wounded. But at present we had no way of retrieving them, we were stranded deep in the Green Zone with no LZ, and we had a strafe coming in on top of us.

I rolled on to my back and searched in the sky to the west. The F-15 would be coming in right over our heads – that's if he'd got his attack run right. Anything else, and we were fucking dead.

As I lay there, I thought momentarily of my wife and young Harry and Ella. God knows I'd miss them, but if I had to die anywhere, at any time, with anyone – then it would be here, in this fight, with these lads all around me. I'd rage and bleed and die for these blokes – every last one of them.

'Tipping in,' came the voice on my TACSAT.

As the pilot spoke, I saw the glinting sliver of a knife-sharp jet arrowing out of the west and into the rising sun. The F-15 was coming in low and fast, and the pilot looked bang-on for the line of attack that I'd given him.

'Visual two pax, crawling forwards on top of your lead platoon's ditch position,' the pilot told me. 'Call for clearance.'

'You're clear hot. Ground commander's initials are SB. Kill 'em!'

The F-15 spat fire. 'Brrrrrrrrrrrrrrrrrrrrrrzzzzzzzzzzt!'

It streaked right over us, the six-barrel cannon flaming, and threading the smoke from those muzzle flashes in a thick trail across the valley. In an instant it was past, the roar of the jet's massive after-burners frying the air, as the pilot put pedal to the metal climbing for altitude.

'BDA!' I yelled. 'BDA!'

'BDA: two dead,' the pilot replied. 'Your lead platoon …'

'Break! Break!' *Dude One Four* crashed in on the traffic. '*Widow Seven Nine*: they're running at you in large numbers from the north-east around to south-west of your position. *Widow Seven Nine*, they're lookin' to fuck you up bad down there.'

'Well get the fuck in and start killing 'em,' I screamed at him. 'Get in and kill 'em, fast!'

'Roger. Attack instructions.'

At that moment, the predatory silence erupted. The bush sparked with muzzle flashes in every direction, as a savage barrage of concerted fire tore down on us, and RPGs flared in the shadows. I didn't need Alan yelling intercepts at me to know what was happening: the enemy commander had given the word for the final attack.

'*Dude One Three*, you have our friendly grid,' I screamed. 'No call signs will move position. Map it out on your computers, get your cannons going and start fucking smashing 'em.'

'Roger. We're lining up for attack runs now, shooter-shooter. Stand by.'

'Dude call signs, you're cleared to attack. Just get your bastard guns going!'

The air above us was alive with the angry snarl of rounds. I felt a sharp tug, as one smashed into my donkey dick aerial, whining off into the undergrowth. If they took out the TACSAT, then we were well and truly buggered. I tried to burrow deeper into the shit and stench of the ditch. I was climbing inside my bloody helmet.

'Brrrrrrrrrrrrrrrrrrrrrrrzzzzzzzzzt!'

The roar and thump of battle was torn apart by a long and beautiful strafe.

'Brrrrrrrrrrrrrrrrrrrrrrrzzzzzzzzzt!'

A second came a few moments later, as *Dude One Four* followed after the lead F-15's attack.

From the skies above the kill zone it was raining chunks of red-hot shrapnel. One landed smack-bang in the middle of the map, and lay there, smoking. I shoved it aside, hardly noticing the burning in my fingers, as I yelled for a BDA.

'BDA: three killed,' the lead pilot confirmed. *Result.* We were starting to smash them back. 'Banking around now.' A beat. 'Engaging.'

The F-15s came in again, shooter-shooter, smashing the bush to either side of us. An instant later a series of agonised, unearthly screams rent the air. Nothing on earth sounds like the cries of a wounded man, especially one torn apart by 20mm cannon fire.

I'd never had a hunger to hear that noise before. But now the enemy were taking casualties and those screams sounded good. The Dude call signs came arrowing in on two further strafing runs, but the enemy fighters just didn't seem to care. They were blind to their casualties and closing in from all sides. The violence of the firefight was numbing.

Along with the lads all around me I was pumping rounds into the bush, aiming with my needle-sight at the flash of movement, or the burst of muzzle flame in the shadows. But as one went down, another took his place. How many of them were there? And how long would our ammo last? And for how long could we keep beating the fuckers off?

The chuntering of some big, nasty weapon joined in the deathfight now. Its deep, throaty thunk-thunk-thunk tore into our thin line of men, rounds smashing apart tree trunks and making mincemeat of the branches above us. It sounded like a Dushka, and the noise of those 12.7mm bullets tearing apart our positions was horrible.

'Tipping in.' *Dude One Three*'s radio call dragged my mind back to the air war. 'Commencing fourth strafing run.' A pause. 'Engaging.'

As the dense funnel of 20mm rounds thrummed through the air above, the second F-15 pilot came up on the radio.

'*Widow Seven Nine, Dude One Four*: sir, this isn't working. They're charging your positions from the north-east, dozens and dozens of 'em. We need to switch to bombs, sir. If not, you're finished.'

'Roger. Stand by.'

God, give me a few moments to fucking think.

The patrol was strung out some two hundred metres from end to end. The enemy were right on top of us, danger-close in all directions. The danger-safe frag distance for a five-hundred-pound bomb – the smallest an F-15 carried – was three hundred metres. Basically, we were fucked if they started dropping bombs. But it was either that, or get captured and killed.

An idea came to me. I'd learned about it in JTAC school, but it was rarely if ever used for real, in combat.

'I'm calling a bug-splat!' I yelled at Chris and the OC.

The two of them ceased firing for an instant, and stared at me, like: *What the fuck's a bug-splat?*

By calling a bug-splat, I was gambling on the exact trajectory of the bombing run. I'd be dropping five-hundred-pound bombs twenty-five metres from our lads, hoping the momentum of the drop would hurl the frag away from us and into the enemy.

'*A bug-splat?* You're fucking cleared!' The OC yelled, without waiting for me to explain. 'Just get the jets in!'

'*Dude One Three*: I'm calling a bug-splat!' I pawed the dirt-spattered map in front of me, double- and triple-checking our positions. 'I want you to drop a GBU-38 twenty-five metres to the north-east of our lead grid, coming in on a 045-degree line of attack. Confirm.'

The pilot repeated the instructions. 'Banking south-east to begin attack run. Stand by.'

As the jet came around in a screaming turn, my heart was in my mouth. I knew if I had the grid one digit out, I'd kill a lot of our guys. Plus if I'd got my map-reading wrong, or misjudged the line of attack, I might kill the enemy but smash the lads at the same time.

The OC's blind faith in me, his JTAC, was humbling. Yet right now, I didn't have a clue whether I was about to save us, or damn us all to hell.

TWENTY ONE
SAVING PRIVATE GRAHAM

I guess Chris didn't know what a bug-splat was either, but he knew I was about to try something completely fucking desperate. As the jet tipped in, he was screaming for all stations to get their bloody heads down.

Above the crack and thump of battle, I had a call coming in on the TACSAT. I guessed it had to be the F-15 pilot seeking final clearance to do the mother of all insane airstrikes. It wasn't. It was his wing.

'*Widow Seven Nine, Dude One Four*: I'm visual with more enemy rushing your lead position. Scores of heavily armed fighters converging on your lead grid.'

'Roger. Stand by. *Dude One Three*: attack now! I need that fucking bomb in! No change friendlies danger-close twenty-five metres from drop – you're clear hot!'

'In hot,' the pilot confirmed. A beat. 'Stores.'

'Thirty seconds! Thirty seconds!' Chris was screaming over the radio, so the lads still fighting could get their bloody heads down.

As the JDAM plummeted earthwards from 20,000 feet, its four V-shaped tail fins steered it into target. The dumb bomb had been rendered smart by the addition of a simple GPS-based homing system. It was upon that I was relying to smash the enemy, and not us.

In the bomb came, a five-hundred-pound blunt-headed warhead the height and breadth of tree trunk. As it screamed down, the snarling wolf-howl of its inrush drowned out even the battle noise. Then the detonation.

The flash was right on top of us, white-hot and searing. The blast wave punched through the trees like a tidal wave, the snarl of the explosion entombing us in a roaring wall of deafening noise. As the blast thundered onwards across the valley, all around me soldiers lifted their heads and their weapons from the dirt. I could see that the lads here were still alive: it was the boys of 2 Platoon who'd been right under the bomb.

'BDA!' I screamed. 'BDA!'

'BDA,' came the pilot's voice. 'Bomb impacted thirty-three metres north-east of your lead platoon. Five enemy pax killed outright.'

YES! GET IN! But we needed confirmation from 2 Platoon that none of them had been smashed.

The OC was yelling for the lads to check in. '*Arsenic Two Zero, Charlie Charlie One*: confirm all OK.'

Silence.

'Repeat: *Arsenic Two Zero, Charlie Charlie One*: confirm all OK!'

Silence again.

'I repeat,' Butsy yelled. '*Arsenic Two Zero*: confirm all OK!'

Chris and the OC got on the net, screaming for the lads to get on the air and respond.

Finally, a voice came up on the net. It was 2 Platoon's radio operator.

'*Arsenic Two Zero*: sorry about that, but we're deafened. We couldn't hear you. Confirm all are A-OK.'

'Dude call signs: friendlies are all OK,' I yelled into the TACSAT. 'Repeat: all A-OK. I want immediate re-attack, same ordnance, advise best target.'

'*Dude Zero Four*: visual enemy pax north-east your lead platoon. In static positions, thirty metres from nearest friendlies.'

'Roger, attack as before: come in on a 045-degree run exactly, no change friendlies.'

'Roger. I'm tipping in with a GBU-12 to hit that position.'

Chris and the OC put out the all-stations warning for the lads to get on their bloody belt buckles. The GBU-12 is an eight-hundred-pound bomb, so bigger than the first drop. It homes in on the hot point of a laser beam, as opposed to being GPS-guided. As *Dude Zero Four* went in to do the drop, his wing would have a laser fired at the target, to guide in the bomb. In theory, laser-guided was more accurate than a JDAM. But it relies upon the person firing the guidance laser, so more could go wrong.

The pilot started his run, as rounds and RPGs slammed into the bush all around us. We had to smash the enemy fast if we were to get our wounded lads out, and the bigger the bomb the more we could kill.

'Call for clearance,' came the pilot's voice.

'No change friendlies,' I yelled. 'Clear hot.'

'In hot,' the pilot confirmed. 'Stores.'

The second warhead came howling through the skies like it was coming in right on top of our position. For an instant I wondered if the pilots had lost their laser spot, or got the grid a digit off. I held my breath, tensing for the impact.

And then the bomb was tearing past and slamming into the earth. The pulse of the blast was more powerful this time, like a giant's sledgehammer smashing through the trees, as eight hundred pounds of high explosives tore the ground and the air asunder. Blasted chunks of shrapnel and rock and shredded branches were spinning through the air, and smashing back down to earth. I yelled for a BDA.

'BDA: the bomb hit in the centre of the enemy mass,' the pilot reported. 'Scores of dead and injured.'

An instant later we got the call from 2 Platoon: again, they were deafened, but still breathing.

'*Widow Seven Nine, Dude One Three*: visual with pax extracting their wounded to the north-east of your lead grid. Plus visual with six-man RPG team approaching your positions.'

'Leave the wounded, smash the RPG team. No change friendlies, same attack run, you choose munition.'

'Roger. Tipping in with a GBU-12. I can see them firing an RPG: get your boys to get their goddamn heads down.'

I cleared the pilot in, and the second GBU-12 was on its way. As the eight hundred pounds of high explosives tore into the earth some thirty metres to the north of us, the lads and I were burrowing into the dirt of the ditch like proverbial fucking rabbits.

A third massive, ear-splitting explosion engulfed us, the violent suck of the detonation tearing the air from my lungs. It left me shell-shocked and reeling, with the blood pounding in my head and ears.

The crushing roar of the bomb was followed an instant later by the crump of secondary explosions, as RPG rounds cooked off in the bush alongside us. A belch of black smoke spat into the sky high above, blocking out the rays of dawn sunlight that filtered through the trees.

'BDA: direct hit,' came the pilot's voice. 'I saw one guy with a backpack of RPGs exploding all over him. Four pax killed. But two stood up two metres from the impact point and ran off northwards. I have no idea how they're still alive.'

'Nice work, *Dude One Three*: keep hitting 'em.'

There was a cry from Alan, 'Enemy commanders are ordering their men to hold their positions, and to press home their attack!' He glanced at Chris and me. 'We have to get out of here! We have to get out of here …'

There would be time to deal with a traumatised terp later. Right now, it was our injured that were on my mind, plus getting us lot out of here without a horrible friendly-fire incident. We were still being hit from all sides, and there had to be more of the enemy we could kill.

'Dude call signs: tell me what you can see,' I yelled. 'I want targets.'

'*Dude One Three*: I see enemy fighters bugging out with wounded to the north of you.'

'*Dude One Four*: roger that. Enemy pax hauling out bodies to your north-east.'

Fucking smashing job. Maybe we were beginning to win this one at last.

'Dude call signs: do not engage enemy with dead or wounded. Find those still firing at us, and smash 'em.'

'Roger. Stand by.'

The F-15s did several more bombing runs, putting in GBU-12s danger-close to the north and south of our column, and pounding the enemy to either side. Each time, I was shitting myself that we'd get it wrong, and kill and injure our own men. But with each bug-splat we threw the blast away from our positions, and into the faces of the enemy.

It was 0630 by now, and we were forty-five minutes into the maddest hour of our entire combat tour. Those two American pilots above us were the very best. Without that pair of F-15s on station a lot of us would no longer be breathing, and I knew it. As for the lead platoon and their injured, they'd have been killed or in the hands of the enemy by now.

'*Widow Seven Nine, Ugly Five Zero*.' I had an Apache checking in to my ROZ. 'I'm two minutes out with the IRT heavy, request LZ.'

We didn't have a fucking LZ. We didn't even have our hands on the wounded, to extract them. But we sure could use that Apache.

'*Ugly Five Zero*, get the heavy to hold off to the western tip of ROZ Suzy, out in the desert. Unsafe to land as we are in GZ, in midst of contact.'

'Roger that.'

'Ugly, this is our lead platoon's grid: 59368219. Can you get your gunship smack-bang above it looking bastard-ugly, to deter the enemy.'

'Roger. Moving over grid as given now.'

'Dude call signs, push up to 5,000 feet. I'm bringing Ugly in low to deter the enemy.'

'Roger. Pushing up to 5,000 feet.'

'What's the heads-up with the LZ?' I called over to Chris, Peachy and the OC.

'There's a big open field to the south-west of Alpha Xray,' Peachy yelled back. 'Get the Chinook down there.'

It was a bloody risky plan, but what else were we to do? I knew that field well, and Qada Kalay was just south of there. If the enemy were out in force at Qada Kalay, they could use their 107mm rockets, or even an RPG to blast the Chinook out of the sky.

Chinooks rarely if ever went into the GZ to extract casualties, and for good reason. Wherever possible the casevac was done in the open desert, where an iron cordon of security could be thrown around the LZ.

'Ugly, ask the heavy if he'll put down in the GZ,' I radioed the Apache. 'We don't have an alternative.'

'Stand by: I'll speak to the pilot.'

'Roger. And Ugly, stay slap-bang where you are above us.'

'Roger that. I'm not moving.'

I could hear the throb of the Apache's rotors thumping through the air, like the reassuring heartbeat of some friendly beast of prey. Above that, the scream-hum of the F-15's jet engines was letting the enemy know that I still had those warplanes on station.

We had three men down, and we had no idea how bad they were. Plus we didn't know how many of the other lads had taken injuries, and were keeping it quiet. It was amazing what damage the boys could take, and remain in the fight alongside their fellow warriors.

With the air stacking up above us, the contact died down to just about nothing. The OC gave orders for all platoons to extract, with 2 Platoon moving back through us as a protective screen. We threw smoke grenades, and under their cover those most-forwards began to fall back.

A circle of figures bent double came hurrying through the trees. For a moment I felt physically sick. They were hefting a poncho between them, weighed down with the body of a man. Carrying that weight in a makeshift stretcher over such terrain in the furnace of the Afghan heat was hellish, let alone doing so after hours of extreme combat. Whoever that young lad was in that poncho, he had to be in a bad, bad way. Or, in the time it had taken us to smash the enemy from the air, one of our injured had died of his wounds ...

'*Widow Seve*n Nine, Ugly.' The Apache pilot's call tore me away from my dark thoughts. 'Visual three pax with weapons fifty metres east and moving in on you.'

'Roger. Stand by.'

I relayed it to the OC, and he told me to hit them.

'Ugly, *Widow Seven Nine*: hit them with 30mm. Nearest friendlies our position fifty metres south-west.'

'Roger. Engaging now.'

Thump-thump-thump-thump ... the 30mm cannon of the Apache's turret spat out a long, twenty-round burst of pinpoint accuracy firepower. I felt the air around us judder and shake as the shells thumped into the bush.

'Ugly: BDA.'

'BDA: two killed, one on his heels and running away from your direction. I've lost him.'

'Top job, Ugly. I want you to do two long strafes to the east of our position, as we extract, to deter any pursuers, on a north to south run.'

'Roger. Standing by.'

I glanced at the OC. He gave me the nod, and put the order out on the net for the last platoon – us lot – to withdraw.

'Ugly, extracting now,' I told the Apache. 'Fire when ready.'

We began to edge our way backwards, following the route we'd come in on. As we did so, the Apache started malleting the positions we'd abandoned. With a wall of 30mm cannon rounds to our backs, we headed west for Alpha Xray.

We reached AX, and I got the Apache to move south over the Helmand River. I wanted it flying low and fast up and down the water, looking very capable of extreme violence. That way, it would provide a block between Qada Kalay and the Chinook, as it went in to extract the casualties.

The OC got a ring of men thrown around the LZ – an open field of thick green crops, fringed with woodlines. With the lads out in force surrounding the landing point, it was about as secure as we could make it. Still, all it would take was one lucky RPG fired from beyond the range of our lads, and the Chinook would be toast.

On the deck our worst casualty, Private David 'Davey' Graham was being worked on by the company medics. He had a blood-stained bandage strapped around his middle, to keep his guts in, where a burst of AK-47 fire had torn into him. Plus he had drip-bags pumping fluid in to replace all the blood and liquids he'd lost.

Eighteen-year-old Davey Graham was a Minimi-gunner, and having that drum-fed light machine gun at the cutting edge of the patrol had made perfect sense. Davey had taken point, leading 2 Platoon into the enemy terrain.

The ambush had come from out of nowhere, at five metres' range. Davey had taken three rounds under the breastplate of his body armour, and as he'd twisted and fallen a fourth had hit him in the backside. The enemy gunman had stepped around a tree to finish Davey off, levelling an AK-47 at his head.

Before he could open fire, the soldier behind Davey had rushed forward and shot the enemy fighter twice, in the face. He'd then grabbed Davey's body under fire, and dragged him back into the safety of the main body of troops.

Amazingly, Private Graham was still conscious as the medics worked on him. He even asked Andy, the press photographer, to shoot pictures of him as they casevac'd him out of there. Andy had one problem. As the ambush was sprung and Davey had been hit, Andy had dived on to the deck, snapping the lens off his top-notch camera.

He'd flung the broken bit of camera kit at the enemy. Then he and John, his fellow reporter, had hunkered down as the bullets and grenades, and then the 20mm strafes and the big bombs had rained down all around them. Somehow, unbelievably, everyone had got out of there alive.

Sergeant Major Peach popped a green smoke grenade, to mark the landing zone, and I cleared the Chinook in to land. The massive helicopter came swooping in, banking hard and low across the river, the unmistakeable thwoop-thwoop-thwoop of the twin rotors beating out a powerful rhythm on the air.

It took nine men to lift Davey's makeshift stretcher, with one holding up the drip. They rushed him out to the Chinook, clambering through flooded irrigation ditches and hauling him over treacherous mudbanks. The two other injured lads had nasty shrapnel wounds, but even so they tried to refuse to leave their mates, and the battlefield. The OC had to order them on to the Chinook. He told them that for today at least their war was over. Once they'd been patched up at Camp Bastion, he'd get them straight back out to the Triangle. We got that Chinook in and out without it being hit, and the wounded en route to the best medical care a British field hospital can offer.

With the heavy in the air, I got the call from the F-15s above.

'*Widow Seven Nine*, *Dude One Three*: low fuel, we're bugging out. Stay safe down there.'

'Roger. And look fellas, absolutely fucking fantastic. You saved our fat arses today, 'cause we were right in the proverbial. Top job.'

'That's what we're here for, *Widow Seven Nine*. It's not us who's down there taking the hits – we're just up above y'all.'

'Aye, and we owe you guys a good few beers.'

'Affirmative,' the pilot laughed. 'Meantime, we'll drink a few for you at KAF.'

KAF is Kandahar Airfield, where the F-15s were based. And from that day on if ever I had an F-15 check into my ROZ, I'd always have a 'how're you doing' passed down for *Widow Seven Nine*, from those two pilots who'd fought above us that day. The pilot of *Dude One Three* was a Captain Tabkurut, or at least that's what his name sounded like. A lot of the 2 MERCIAN lads – not to mention us lot – owe him and his wing our lives.

An hour later we were back at PB Sandford. *Ugly Five Zero* had stayed with us all the way up Route Crow, shadowing us in until the gates clanged shut on the last man of the patrol.

I glanced around me. There were lads everywhere slumped against the walls, wiping the sweat and shit and blood off their faces. Every man was a picture of shattered exhaustion. The looks in the eyes said it all: *How the hell did we get out of that lot?* With the last of their energy, lads struggled out of body armour and helmets. The adrenaline was pissing out of our veins now, to be replaced by a crushing, leaden fatigue. The OC had ordered all men on foot patrols to wear full body armour and helmets: today's action had proven his decision 100 per cent the right one.

Had the bullets that had hit Davey Graham been an inch or two higher, the breastplate would have saved him. As it was, the enemy gunner had sneaked the rounds in beneath the lower edge, tearing apart Davey's guts.

The newly-qualified 2 MERCIAN medic had done the emergency first aid on him. But in spite of her best efforts, barely a soldier amongst us doubted that Davey Graham was a dead man.

Private Davey Graham was a fresh-faced lad with a ready smile and a pair of piercing blue eyes. He was known as being a bit of a joker. Earlier in the morning I'd seen him holding up his helmet on the end of a stick, in a mock gesture to draw enemy fire. Graham was the kind of bloke you'd have trusted with your life, and he'd more than likely volunteered to take point on the lead platoon. After the shock of fighting so hard to save him, the idea that we could lose him was hitting us all bastard hard.

I loosened my own body armour, and went to heave it over my head. As I did so, I felt a stabbing, jabbing pain deep in my left shoulder. It was only now that I remembered the violent thump to my back that had sent me flying face-first into the ditch.

I craned my head around, and I could just about see this huge spreading purple-red bruise where my left shoulder met my neck. Something big must have hit me, and cannoned off the top of my body armour. I didn't dwell on it for long, or breathe a word to anyone. I was alive and in one piece: others hadn't been so lucky.

In any case, there was barely a man amongst us who hadn't taken a lump of frag here or there, or a blast of flying rock and grit, in their body armour. Until those jets had got to work hitting the enemy with their five-hundred- and eight-hundred-pounders, we'd taken a right malleting.

A couple of the 2 MERCIAN lads came up to me.

'Nice one, like, Bommer,' said the one.

'Yeah, cheers and all that,' said the other.

'Cheers for what?' I asked.

'For saving us necks out there,' one said.

'With the airstrikes and stuff,' said the other.

'Aye, well, get the kettle on, will you lads. Time for a brew.'

As they wandered off, I heard one say to the other: 'We got second place today. Runner-up prize. Bommer's right – better get the tea on.'

It was 1030 when the OC gathered us together for a chat. I'd seen him giving a couple of the lads a fatherly slap and a hug, and I knew we were all feeling it.

'I've just had word that Davey Graham made it back to Camp

Bastion alive,' the OC announced. 'He's been badly shot up, and will be evacuated to the UK just as soon as that is possible. He's in a very serious condition, and will need to be operated on. But we got him out of a massive enemy ambush, and we got him back to Camp Bastion. And for that, every man amongst you should feel justifiably proud.'

We drifted off, each trying to find our own little patch of personal space. Not easy to do, in a mud-walled compound crammed full of fifty-odd soldiers. In spite of the OC's words, we knew that we'd been smashed. The enemy had had the upper hand, and it was only the air, and a shedload of good luck, that had saved us.

I grabbed a brew before the rush, and headed for the stairs leading up to JTAC Central. I knew I could get some headspace up there. At the ammo-box staircase I got collared by the OC. He fixed me with a look, and for a second or two he said nothing.

Then: 'Cheers, Bommer.'

That was all. Butsy was the type of bloke who didn't give praise easily. But you knew from the expression in his eyes if he was happy or not, and by Christ his eyes had said it all. *Cheers Bommer.* That was enough for me.

Up on the roof I took a slurp of my brew. I'd ladled in the sugar to give me some energy. All I kept thinking was this: *Please do not give me any more air.* All I wanted to do was to get on the blower and speak to Nicola. Today was our wedding anniversary, and I'd not been able to wish her a happy anniversary.

Down below I could hear this thick Cockney voice going: 'Fucking 'ell! Fucking 'ell!' Over and over the same phrase, in turboclip mode. It was Andy, the press photographer. He and his reporter mate kept laughing and laughing. They'd been with 2 Platoon in the heart of the ambush, and they were fried.

'Fucking 'ell,' Andy kept repeating. 'I can't believe that's what you guys go through every day. Fucking 'ell.'

'You're fucking lucky, lad,' 'Mortar' Jim, 2 Platoon's mortar-operator replied. 'That's the worst we've ever had it.'

I finished my brew, came down from the roof, and got on the satphone to Nicola. I wished her a happy wedding anniversary, and asked her what she was doing for the day. She told me she was having a nice meal with the nippers, Harry and Ella.

'What's your day been like?' she asked me.

'Well, nowt much's been happening,' I lied. 'We've had a bit of a boring one.'

'Paul, what's wrong?' she asked. 'You don't sound like you normally do. What's wrong with you?'

'Nowt's the matter, love. It's just, I was up at the crack of dawn and I'm well-knackered.'

After the call, I went and joined Throp and Sticky in the Vector. Sticky was staring ahead with that wired, 'thousand-yard stare' look that I guess we all had to have by now. It was the weird, unfocused, shell-shocked look of having been in the fight of your lives for hours and hours on end, not to mention the weeks of combat before.

'Is it all ever worth it?' Sticky muttered.

'Is what ever worth it, mate?' I asked.

'Any one of us could've got whacked from those bombs you called in.'

I shrugged. 'Aye. Top bloody present that would've been on the wife's wedding anniversary.'

'So, is it ever worth it, for eighteen hundred quid a month?'

The only answer was a chorus of Andy's 'Fucking 'ells' that drifted across to the Vector.

TWENTY TWO
BIN LADEN'S SUMMERHOUSE

At stand-to the following morning I had two Apaches – *Ugly Five Zero* and *Ugly Five One* – check in. The pilots were learning that ROZ Suzy was seriously busy. They'd fly at higher altitude in an effort to conserve fuel, so they could offer me a few minutes' play-time en route back from whatever mission they'd been on.

The Apache pilots were gutted that a major action like yesterday's could have gone down without Ugly playing a bigger part in it. For once I was glad to have had those jets over us, as opposed to Apache. Only something with the capacity to drop serious ordnance could have beaten off an attack of the ferocity that we had faced.

I got the Apache pilots searching over the positions of the previous day's battle. But apart from a smoking cooking fire at Golf Bravo Nine One, there wasn't a sign of life anywhere.

A couple of days went by with only sporadic attacks against us. The odd burst of small arms fire and 107mm rocket barrages hit PB Sandford, but there was nothing resembling a full-blown attack.

Alpha Xray got malleted from the woodline at Golf Bravo Nine One. I couldn't get any air, so Chris called in a barrage from the 105mm howitzers and drove the enemy off. It was like they were probing us all over again, in an effort to test our resolve and our lines of defence.

Golf Bravo Nine One was fast becoming the enemy's start line for any assault. Their headquarters we reckoned was back at Golf Bravo Nine Eight – the position that our foot patrol had stumbled into. That would explain why they had fought so ferociously, throwing in waves of fighters in an effort to annihilate us.

We pushed a patrol down to Alpha Xray, on foot and with two Snatch Land Rovers. Throp and I went on it, in part to defuse the tension of being cooped up in PB Sandford, and in part 'cause I had air over the convoy. I got the Dude call signs flying recces to the east of AX, around where Davey Graham had been gunned down.

Nothing was seen. Throp and I tabbed back towards PB Sandford, along with the platoon from AX that had been relieved. As we did I lost the air. The F-15s were ripped to a TIC somewhere else in Helmand. The radio chatter was going wild that they had eyes on the patrol, but even with the F-15s gone there still wasn't a sniff from the enemy.

That night I got a pair of A-10s above me. We'd got Intel that the enemy were doing a major resupply by vehicles out in the desert. It was all part of their build-up in the Triangle, the ultimate aim of which was to smash us. Intel reports had eight or nine vehicles involved in the resupply. In due course the A-10 pilots found a desert convoy.

Via my Rover terminal I could see the group of vehicles the Hog call signs had discovered. But as Chris, Throp, Sticky and I studied the images, we couldn't see anything that resembled ammo or weapons. For all we knew it could be a midnight wedding. We decided we couldn't hit the convoy, and we let it go on its way.

By morning, the radio chatter was hot about a successful resupply. Enemy units were being ordered to fetch new weapons and ammo. I had a pair of F-15s overhead, and got them flying air recces all across the Triangle. But not a thing was moving down there, not even farmers working their fields. I'd never seen it this quiet. It was weird. Spooky.

In desperation, I got the F-15s to fly search transects over the old Soviet trench system, in the desert four kilometres to the north-east of us. There was more than a kilometre of interlinked earthworks, where you could move from position to position without being seen.

We reckoned those trenches linked into an underground tunnel system, stretching all across the Triangle. How else could the enemy resupply their fighters, without being seen from the air? We'd had reports that the Triangle was honeycombed with hidden caverns and tunnels, and the body of evidence was growing by the day.

A couple of days back I'd had a Predator over the Green Zone. As it had flown its recces, the drone had passed over this small, tower-like building, enclosed on all sides by thick woodland. There were four males visible, one on each corner of the roof. I couldn't see any weapons, but those guys sure looked like sentries to me.

The building was some 2.5 kilometres east of Alpha Xray, so well into enemy territory. It was way beyond the Golf Bravo codenames, and into the Golf Charlies. I got the Predator to loiter over that grid. I saw a figure leave the building and walk along a path for a minute or so. He reached the middle of a field and completely disappeared. One moment he was there, the next gone.

I watched another male of fighting age leave the building, follow the path to the centre of the field, and puff – he was gone. By the third time, I was convinced I'd found the entrance to a tunnel system. More than likely it had been built during the time of the war against the Soviet Red Army, and would lead all the way back to the Soviet trenches.

The building in the woods was in a perfect defensive position. It sat in a crook of the Helmand River, on a promontory. It was invisible from the ground, being surrounded by thick woodland. It was only by luck that I'd spotted it from the air. I nicknamed it 'Bin Laden's Summerhouse', and the name just stuck. Word spread, and I started having pilots ask me if I'd spotted Bin Laden himself there yet.

I got another Predator in overwatch of the Summerhouse. This time, there were fifteen males sat under the trees, getting briefed by a guy leaning on a motorcycle. Not one of them was showing any weapons, and I'd yet to see a sniff of a gun. But my instinct was screaming at me that this was a major enemy hub.

The guy finished talking, got astride the bike, and was driven off down the track by his 'driver'. I tracked them for fifteen kilometres moving in the direction of Sangin. En route they kept getting waved through by groups of males of fighting age. Finally, they reached a crossing of the Helmand River and boarded a boat. Whilst on the water the main figure swapped his black turban – the uniform of the Taliban – for a white one. Around about then I lost the Predator. But I'd bet any money that the guy was some Taliban bigwig, and the Summerhouse some high-level enemy base.

Of course, everyone from the OC down wanted to go in and hit the Summerhouse. But it was a good kilometre beyond Golf Bravo Nine Eight, the point at which we'd walked into the Davey Graham ambush. It would take a lot more blokes, and a lot more firepower, to battle through to there.

I kept the memory sticks of all the material that I'd recorded from the Predator feeds. I passed the lot up to Nick the Stick, and those in command of his group of elite American warriors. It was better to leave it up to those boys to hit the Summerhouse, with maybe a 'Spooky' call sign and some Apaches on hand to assist. For now I had my own priorities to deal with. Chief of those was trying to work out where the hell the entire human presence in the Triangle had got to. Maybe the enemy were down in their tunnel systems, sorting out their ammo resupply, and briefing their newly arrived fighters. That would fit with the Intel that was coming in.

We decided to take advantage of the enemy going to ground. We headed out with the Czech Army unit, in their Toyota wagons encased in camo-netting and mock-greenery. Watching from a distance, the Czechs on patrol looked like a line of moving bushes, albeit with long-nosed Dushka heavy machine guns poking out.

We drove past Alpha Xray and pushed on to Golf Bravo Nine One. As there was nothing doing with the enemy, we wrapped a couple of strings of plastic explosive around each of the trees, and blew a long line of them sky-high. Kaboom! Kaboom! Kaboom! It wasn't that we hated trees: we just didn't like the cover they provided for the enemy. Golf Bravo Nine One was the front line from where they kept hitting Alpha Xray. By blowing the treelines, we deprived them of cover via which to sneak up on our base. Alpha Xray was getting whacked pretty much on a daily basis, and we didn't want to make it any easier for the bastards.

The following morning Alpha Xray got hit just after first light. Maybe we'd needled the Taliban by blowing up their greenery. Either way, there was a barrage of small arms and RPGs smashing into the base. I couldn't get any air, so Chris called in fire, pounding the enemy with the 105mm guns and our own 81mm mortars.

But none of this was the big one, and we knew it. Something nasty was brewing. We kept having walk-ins warn us that the enemy were reinforcing and rearming for a big push. They planned to overrun one of our bases, and we guessed it had to be Alpha Xray.

The OC decided to push out a foot patrol to the east of Monkey One Echo, and into the Golf Charlie codenames. The aim was to get a rise out of the enemy around Bin Laden's Summerhouse, so we could pre-empt their big push by forcing them into a fight. We reckoned the Summerhouse was 'the Mosque' that the enemy commander kept calling their men to for pre-battle briefings. If the Summerhouse and the Mosque were the same place, and our foot patrol could provoke them into opening fire, we could smash it from the air.

We pushed east and crossed the enemy front lines, creeping deep into their territory. I had two A-10 Warthogs in support. I had one watching to the front of our patrol, and the other with eyes on the Summerhouse/Mosque. We hit the Golf Charlie One Seven area, and a lone RPG went sailing over our heads. It smashed into the bush twenty metres beyond us. No one could see the firing point, and apart from that it was dead all around us. Not a soul was to be seen.

We got back to PB Sandford without another shot being fired, the A-10s shadowing us into base. It all confirmed what our walk-in sources were telling us: the enemy had pulled back to resupply and rearm, in preparation for the big one.

The following day was Born Naked Day, or at least it was for the Czech Army unit. The Czechs had claimed their own corner of PB Sandford, where they kept their Mad Max Toyotas parked up between two massive mud ramparts, like blast walls.

It was Saturday, 11 August and the Czechs intended to spend the entire day naked, no matter what. They'd bloody fight naked if they had to, or so they told us.

They erected a sign at the entrance to their domain:

> The World Famous Czech Born Naked Day.
> Make Love Not War.
> All proceeds to the Children of Chernobyl Fund.

Then there was a list of rules.

> Get Naked.
> Stay Naked.
> Extreme nudity.
> No clothing ever.

It wasn't exactly our sense of humour: that was more of the Get Snoopy kind. But maybe we were just repressed when it came to getting our kit off. And for sure we had some bizarre traditions of our own in Britain – like chasing cheeses down mountains, or peashooter contests, or bog-snorkelling. In comparison, Born Naked Day was a no-brainer, especially if there were some Czech girls involved – which I guess there would be when doing it back home in the Czech Republic.

The Czech unit were a massive bunch of lads. They made Throp look positively weedy. Each looked as if he'd been fed on an intravenous drip of steroids during infanthood. If they wanted to sit around with their tackle hanging out, none of us were going to argue. Anyway, it was all for a good cause. We dug deep in our pockets and chucked a load of our hard-earned spends into the Born Naked Day bucket, trying not to get an eyeful of any Czech tackle. One look could give you a serious inferiority complex.

At midday we had an Afghan elder walk-in. We steered him away from the naked Czech monsters, and grabbed Alan, our terp. The word from the elder was that the enemy were moving back into the Triangle in big numbers. They were reinforced and rearmed, and their intention was to hit us hard and drive us out of here.

The OC decided to pre-empt them. He was a man who believed in fighting on the terrain and at the time of his own choosing. We'd push out a patrol on foot past Alpha Xray, then hook north parallel to Route Buzzard, probing north of the area in an effort to force the enemy's hand.

The morning of the patrol I had a bit of a problem. I didn't know about it until I broke wind, and then I had it all down my legs and in my boots. There were several cases of diarrhoea and vomiting in the base, and now I'd got a dose. I had two more bouts that morning, and eventually I had to accept that I wasn't going anywhere.

I perched in the back of the Vector, from where I was in sprinting distance of the shitters. I had another attack, but failed to make the bogs in time. I stripped off and crouched in the wagon in my undies and flip-flops. It wasn't a pretty sight and I smelled rank, but I was more concerned about controlling the air, and making sure the lads on the patrol were all right.

They left PB Sandford at 1345, and pushed down Route Crow towards Alpha Xray. Just as they neared the base there was an eruption of small arms and RPG fire. The enemy were in the sawn-off trees at Golf Bravo Nine One. I got on the air and requested immediate Close Air Support (CAS).

As I did so, there was a massive, punching blast from the direction of Monkey One Echo. I glanced at Sticky, who'd opted to stay with his JTAC, despite the fact that I'd shit myself. He was hanging by the door of the wagon, where I guess the smell was a little less lethal. He was on his radio immediately, calling for a sitrep from MOE.

'They got a man down!' Sticky relayed to me. 'A lad's been hit by an RPG!'

'*Widow TOC, Widow Seven Nine*. Sitrep: we're under two-pronged assault and we've got a man down. We need IRT stood up right away.'

'*Widow Seven Nine, Widow TOC*. Roger that: stand by.'

My mind was fucking racing. If a lad had taken a direct hit from an RPG, he was more than likely spread across several acres of desert. So we'd more than likely lost another one. I felt the rage sweeping over me. In spite of my compromised state, I felt this irresistible urge to grab my SA80 and go out and smash some enemy. Even from inside the Vector, the crack and thump of battle from both our bases was deafening.

'*Widow Seven Nine, Widow TOC*. *Recoil Four One* and *Recoil Four Two* are inbound into your ROZ. *Ugly Five Four* is bringing in the IRT heavy.'

The call from Widow control brought me back to my senses. I was hardly in a fit state to go out fighting. I got Sticky to get the casualty in to PB Sandford. We'd then do the casevac from an LZ just to the north of the base in the open desert. The Harriers came on station, and I got them flying an immediate low-level show of force, screaming over the walls of Monkey One Echo.

The casualty was loaded into a Vector, which hurtled across the high ground to us. The RPG had impacted a metre away from the injured lad. It had ploughed into the wall before exploding, which had kept the frag down. He was in a bad way, but the medics reckoned we could save him, as long as the Chinook got him out in time.

I got Apache *Ugly Five Four* inbound with the Chinook. I got the Harriers banked up high, so I could get the heavy in. We got the massive helicopter down on the LZ, the injured lad was rushed aboard, and then the Chinook was on its way again.

Ugly Five Four stayed with me, and I got it hunting for that bastard RPG team. But it was like looking for a needle in a haystack. As soon as the Apache was overhead, it all went dead quiet. On the airwaves enemy commanders were ordering their men to: 'Remain in ambush positions, then the helicopters won't see you.'

The lads came back from patrol, but once they heard that we'd had one smashed by an RPG they went wild. They wanted to get right back into the Green Zone in full battle rattle, and find the enemy. They calmed down a bit when they learned that the medics reckoned he'd make it through all right.

I got allocated an Ugly call sign for the following morning. It was unprecedented to be given an Apache without a TIC. I soon found out why I'd got it. We had a convoy coming in on a resupply, and there was a Sky News team on it.

The Apache was to shadow them in.

TWENTY THREE
THUNDERBOX MAYHEM

Every time you went for a piss you had to do ten pull-ups. That was the law. I reckoned it was a bit harsh on me, 'cause I drank so much tea, but there it was. All the lads at PB Sandford had agreed to it, and woe betide anyone who shirked.

We'd already been hit by something that morning – maybe a 107mm barrage from Qada Kalay. We'd been in the middle of a cricket-off, which the Sky crew were filming. Suddenly, there was a yell of 'Incoming!', and everyone went scrambling for body armour and helmets, not to mention some cover.

Now we were at the pull-up bar. We'd cobbled together a gym from old ammo crates, steel pickets and black nasty. It did the job. The pull-up bar was a beam slung between two walls. It was getting a bit competitive, as we had the Sky cameraman filming us.

The pull-up king in the FST was Chris. He wasn't called Johnny Bravo for nothing. He'd just managed twenty-four. Sticky had followed with a miserly eighteen. The most I'd ever done was twenty-two, but I was determined to beat Sticky. I got on the bar and reached sixteen, when suddenly the lads started going wild.

'Come on, Bommer! Come on, Bommer!'

'You can do one!'

'Come on, you fat fucker!'

I did another – seventeen – and the lads were going crazy. They were doing just about anything they could think of to make me laugh. I was trying not to, but as I went for number eighteen – making me evens with Sticky – I lost it. I gave up and dropped off the bar.

I turned around to find the Sky cameraman filming me. He'd only started shooting when I was on number sixteen, so it looked as if I'd managed just the one pull-up. I have a giant Angel tattooed across my shoulders, and it isn't exactly a common tattoo. Anyone seeing that on the news would know instantly that it was me. The Sky cameraman was pissing himself. The lads were laughing their rocks off. I waited until they'd calmed down a bit, then tried giving the Sky bloke one of my looks. But I couldn't help the silly grin that kept twitching at the corners of my mouth.

'If you put that on Sky News, I'll batter you and hand you over to the Taliban,' I told him.

We cut a deal that the footage would never be shown, as long as I pulled in some class airstrikes whilst the Sky crew were with us.

Sadly, that morning we were losing the OC. The resupply convoy was setting off for FOB Price, taking Major Butt with it. He was devastated to be leaving his lads, and before the end of their tour, but orders were orders. He'd done his allotted stretch in command, and a new guy was taking over.

It was hard to see Butsy go: he'd been like a father to us. He did a little speech, which was all about how he didn't want to leave after all we had been through together. But he had to let the new OC come in and do his job. Major Butt had been fucking brilliant. He was a fantastic OC. As he spoke, a few of the lads were close to tears.

I was gutted to see Butsy go, but I warmed to the new OC quickly. Major Stewart Hill was a tall, dark-haired rake of a guy, and he was to prove himself to be a top bloke. He was the kind of officer who wouldn't ask his lads to do anything he wouldn't do – a bit like Butsy, really. We couldn't have wished for a better replacement in the new OC.

We got orders from Major Hill to push a patrol down to Alpha Xray. The Sky team were going to try a night down at the Alamo. Much that I might want the cameraman to get captured and that tape destroyed, I reckoned I might need some air on hand. After all, we were going to have a High Value Target (HVT) – the Sky cameraman and correspondent – deep in bandit country.

I was told that no aircraft could be spared, not unless we had a TIC. So I put a call in to *Widow Eight Seven*, a fellow British JTAC who was the nearest to me in the area. A new base – PB Arnhem – had been established to the south-west of us. *Widow Eight Seven* was the JTAC there, and most days we'd have a chat on the air about what we'd been up to. He told me that he had a pair of F-15s flying air recces for him, but that nothing much was happening. PB Arnhem was about seven kilometres away. If I needed air, I could borrow *Widow Eight Seven*'s F-15s.

At 1745 the patrol set out. There were sixteen-odd lads, plus the Sky cameraman and the correspondent, Alex Crawford, who looked vaguely familiar from the news. They crept down Route Crow into the dusk valley. A hundred metres short of Alpha Xray, the darkened bush exploded in gunfire and the fiery trails of RPG rounds. In seconds, the patrol was pinned down and deep in the shit. They were being hit from fire points all along Golf Bravo Nine One, the enemy's favourite point of attack. I radioed my fellow JTAC, to see if I could rustle up some air.

'*Widow Eight Seven*, *Widow Seven Nine*: patrol in contact. Can I borrow one of your Dude call signs?'

''Course you can, mate,' the JTAC replied. '*Dude Zero Seven*, *Widow Seven Nine* is now your controlling ground station.'

'Roger that,' came the pilot's reply. '*Widow Seven Nine*, *Dude Zero Seven*: what can I do for you, sir?'

'I've got a patrol down at Alpha Xray in heavy contact. Enemy are firing from treelines running south to north and west to east, forming a cross at Golf Bravo Nine One.'

'I know your area well, sir,' the pilot replied. 'I was with you a couple of days back dropping danger-close to you boys.'

It was one of the pilots from the Saving Private Graham patrol. It was good to have him with us again.

'I want that enemy position strafed with 20mm, on a north to south run,' I told him.

'Roger that. Inbound two minutes. Stand by for sixty-seconds call.'

I double-checked my coordinates on the maps. It was a couple of days since I'd controlled a jet doing a live drop on live targets. Happy that all was as it should be, I turned to the new OC who was standing at the Vector's open door.

'Sir, I've got an F-15 coming in to strafe the treeline to the north-east of the patrol.'

'Happy with that,' the boss confirmed.

'Prepare to give the sixty-seconds call, sir.'

I was asking him to put out the all-stations sixty-seconds warning. For an instant the OC stared at me, as if to say – *How on earth have you managed that?* It was barely three minutes since the start of the contact, and CAS would never normally be with you in under fifteen. There was no time to explain.

The pilot came up on the TACSAT. 'Sixty seconds out.'

'Give an all-stations sixty-seconds warning, sir,' I repeated. Then I cleared the pilot in. 'Friendlies one-twenty metres west of target. You're clear hot.'

'Tipping in.' A beat. 'Engaging.'

'Bbbbrrrrrrrrrrrrrrrrrrrrrrrrrrrrrrrrrrrrrrrrrrrzzzt.'

The strafe echoed across the darkened valley, as the 20mm rounds hammered the north–south woodstrip. Just as soon as the roar of the gunfire had died away, I got the pilot to bank around and do a second strafe to hit the west to east treeline.

'Tipping in,' the pilot confirmed. 'Visual two pax with weapons in woodline.'

'Hit 'em,' I told him. 'No change friendlies. Clear hot.'

A second burst of cannon fire echoed across the Green Zone.

'BDA: two dead,' the pilot confirmed.

I banked him around again, and did a third run of 20mm. Then I asked him to do a recce of the terrain at Golf Bravo Nine One.

'Visual three pax with weapons and muzzle flashes in a ditch just north of Golf Bravo Nine One,' the pilot told me.

'Bank west, and hit them with a GBU-38.'

I warned Chris and the OC that I was dropping a five-hundred-pounder, and for the patrol to get their heads down. It was a hundred and twenty metres danger-close at night, and a couple of months back I wouldn't have dreamed of doing such a drop. But I had my favourite pilot above me, and this was the only way to fight the enemy in Helmand.

'Tipping in,' the pilot radioed.

'No change friendlies. Clear hot.'

'In hot.' A beat. 'Stores.'

In the thirty seconds it took for the smart bomb to come snarling down on us, I had Chris and the OC yelling over the net for the lads to get on their bloody belt buckles. A hollow thump ripped apart the night as the bomb hit, the white heat of the blast throwing angry red shadows across the walls around the Vector.

'BDA: three pax dead. No further movement around Golf Bravo Nine One.'

The contact had died down to nothing. I thanked the F-15 pilot, and pushed him back across the river to *Widow Eight Seven*. I got on the air and sent a sitrep to Damo Martin, in the FOB Price air-planning cell.

'I've just done three 20mm strafing runs, and dropped one GBU-38,' I reported. 'At least five Taliban killed.'

'Good work,' Damo replied.

Then this: '*Widow Seven Nine, Widow TOC*: what platforms were you using?'

'I pinched a Dude call sign from *Widow Eight Seven*.'

'Roger. Wait out. Stay on these means.'

The duty officer at Widow TOC was asking me to stay on this frequency. He'd sounded a bit confused. Maybe even annoyed. A minute later he was back.

'*Widow Seven Nine*, what were you doing pinching aircraft?'

'We had an HVT out on patrol and pinned down in the Green Zone. I needed air.'

'There's a set procedure for getting air. You've broken every rule you could have broken, as we knew nothing about the contact.'

'Well, it doesn't matter, 'cause there's no contact any more: I've killed them all.'

'You are not – I repeat not – to do that again.' The guy on the other end was fuming. '*Widow Seven Nine*, you are to stick to set procedures.'

That was that. End of my bollocking over the air.

That F-15 control was the fastest I'd ever done in theatre. It was eight minutes, from start to finish. I got a brew on, and then we got a call from Alpha Xray. The patrol had got safely in to the base, and the Sky team were bird-happy. They'd got the whole of the contact on film, and were wowed by the speed and power of the airstrikes. Needless to say, the lads under siege at AX were chuffed as nuts too.

Stewart Hill came and found me. He was fuming at what had happened with Widow TOC. As far as he was concerned, we'd had a patrol plus HVT under attack, and his JTAC had pulled a blinding move to relieve them. And I'd been bawled out for doing so. I appreciated the OC's support. He was clearly 100 per cent there for his lads. But I didn't really give a shit that Widow TOC had chewed me out over the air.

I had a brew and a fag, and got to bed with Alpha Xray safe as houses, and the Sky crew well happy. What could be better than that? Enough said.

After stand-to the following morning the patrol returned with the Sky crew in tow. Sticky and I were having a chat, when I felt a rumbling in my stomach. I still wasn't right after my attack of the runs. I warned Sticky I was off for a crap, and made a dash for the shitters.

The thunderboxes were a pretty basic affair – a plywood wall wrapped around with HESCO, with holes cut in a plank bench to do your business. The HESCO was shoulder-high, so you could sit there having a crap and chat to your mates outside.

Being a bit of a petrol-head, I'd grabbed a copy of *Auto Trader*. I was looking forward to having a good read whilst I was on the throne. I sat on the middle of the three holes, and buried my head in the magazine.

A few moments later I noticed a figure coming towards me from the main compound. It was the Sky reporter, Alex Crawford, and for a moment I was a bit embarrassed. But I thought: *I'm only having a crap, and there's nowt wrong with that.*

'Morning,' I said.

'Morning,' she replied.

Then she walked in, pulled down her hoggers and perched on the hole next to me. There were eight inches on my left separating us, as she proceeded to have a dump right next to me. I had my feet on a sandbag and my combats around my ankles, and I tried desperately to bury my head deeper in the magazine.

She started going on about what a fantastic job I'd done with the air the night before. I felt so awkward I didn't know what to say. I tried to quieten down my doings, but there was a sudden breech explosion and I let rip. It was like you'd do after having a kebab with chilli sauce and after a night out on the beers.

I could feel myself going bright red in the face. I stuck the *Auto Trader* higher in an effort to hide my discomfiture. As soon as I could I finished off. I was out of there like a shot, leaving Alex Crawford alone on her throne. No way could I talk to her whilst we were both having a dump and only eight inches separating us.

I went to the chill-out room and threw the *Auto Trader* on the pile, then made for the privacy of the Vector. Major Hill was standing at the wagon's open door.

'Fuck me, Alex Crawford has just had a dump right next to me on the thunderboxes,' I remarked. 'And I mean, it's just not ladylike.'

The OC cracked up laughing. 'Bommer, she reports from all over the Middle East. She's hardly going to stand on ceremony if she needs a quick crap, is she?'

'It wouldn't have been so bad if it was just a dump she was having,' I complained. 'It was the way she was trying to have a cosy chat about my air from the night before. Think about it,' I went on. 'Some bird you've seen on the telly sits next to you and starts having a dump – well, it's fucking weird.'

The OC and the rest of the lads were killing themselves.

A few minutes later Alex Crawford wandered past the wagon. I could feel myself going red as a beetroot. Chris, Throp, Sticky and Jess were torturing me. Three hours later they were still winding me up, and I was still hiding in the wagon.

All of a sudden there was a yell from Mikey Wallis, followed by a long, deafening: 'BWAAAAAAARP!'

Mikey was giving a blast on the air horn, meaning there was a mortar round in the air. Mikey Wallis had the worst job in the Triangle. He sat in a bunker about the size and shape of a toilet cubicle, staring into his mortar-locating radar screen all day long. When a round went up he got his moment of glory, and punched the air horn. He was doing a blinding job of it too.

The lads ran around grabbing body armour and helmets and taking cover. Alex Crawford started to film what looked like a live report. She was stood before the cameraman in blue helmet and matching body armour, looking very much the part. I shook my head: I doubted if I'd ever be able to watch Sky News again without blushing.

I can't remember where the mortar landed. We were getting hit on such a regular basis I'd given up noticing. Anyway, the Sky crew must've decided they had enough in the can by now, for they moved off with the convoy back to FOB Price. I wasn't overly sad to see them go. At least it meant I could stop hiding in the back of the Vector.

Later, I rang home and spoke to the wife. She sounded unusually excited, and it turned out I'd been spotted on TV. They'd been watching Sky News, and all of a sudden there was a report from our base. There was a scene of me in the back of the wagon in my shorts and T-shirt, on the TACSAT talking to some air.

Harry had rushed forward and pointed at the screen: 'Look! Look! There's Daddy!'

I didn't tell Nicola about my toilet troubles of earlier. I preferred the image of me in the back of the Vector looking manly, to me and Alex having a cosy chat in the shitters.

We had two new developments on the Intel front, but it was hard to assess how real they were. We had Intel from elders who'd approached a patrol. Word was that the enemy had brought in a seriously big weapon with which to hit us. There was no telling exactly what it was, but it was a step up from what we were used to.

The second lump of Intel came to me via a spy plane. The aircraft had picked up enemy comms about a new commander arriving in the Triangle. Commander Hadin was a nephew of Commander Jamali – the previous big cheese and the cop killer – the guy I'd smashed with the Harrier, putting a JDAM through the roof of his 'hardened position'.

We had no idea how accurate the first piece of Intel was. But apparently, Commander Hadin was getting his men in position and 'was ready to attack soon'.

In which case, the OC decided we'd go and pick a scrap with them first.

TWENTY FOUR
WOLF MAN

Like all good battle plans, this one was simplicity itself. We'd take a big patrol down to Alpha Xray. We'd do whatever it took to get a rise out of the enemy at Golf Bravo Nine One, their front line of defence. Once they'd revealed themselves, I'd flatten them from the air.

We went out in provocative strength. The patrol consisted of the new OC, his HQ element, our entire five-man FST, plus two platoons. By the time we'd reached Alpha Xray we'd not had a sniff from the enemy. It was dark, and we decided to try a new ruse on their fighters over at Golf Bravo Nine One.

We got a sound commander set up on the roof of Alpha Xray. The sound commander is basically an enormous speaker – the sort of thing you'd have stacks of at a Motörhead concert to pump up the volume and blow your eardrums. It sits on its own legs, and has a microphone into which you can talk.

I had one eye glued to the Rover terminal, which was feeding me images from the air. I could see the positions all around Golf Bravo Nine One picked out in the eerie green glow of the aircraft's infrared scanners, but not the hint of a glowing human figure could be seen.

I spoke into the microphone: 'Aye-up, Talitubbies, it's us lot here.'

My voice boomed out over the darkened, silent landscape, echoing back to us. All around the battle-scarred rooftop figures hunched over their weapons – SA80s, grenade launchers, Gimpys and the big 50-cal. I handed the microphone to Alan, our terp, so he could translate as best he could what I was saying.

Over the days and weeks Alan had become more and more like one of the team. As Sticky's beard had thickened, so he and Alan had ended up looking like identical twins. I'd made the two of them stand shoulder-to-shoulder whilst I took a photo. When Sticky saw it he had to admit that they did look like two grinning idiot-brothers.

The photo got passed around a bit, and Alan had got renamed 'Sticky's Brother'. That's what we all knew him as now. I had no idea how he was going to take to the sound commander ruse. But we knew how much he hated the Taliban, so we had to presume that he'd like it. So far he was doing a cracking job of translating my words.

'Now, here's a thing,' I continued. 'You might not know this, but Hadin, your new commander, he's on our payroll. That's how we keep tracking you down with them big bombs. Hadin's one of ours. He's on our side. So what d'you reckon to that, Talitubbies?'

I stopped talking, and Sticky's Bro translated. The echoes rolled in from the shadows and the darkness. The silence that followed was broken by a distant cry – the first bellowed response from the enemy. Golf Bravo Nine One was less than two hundred metres away, and the voice drifted over to us faintly on the still night air.

'British soldier! First we fuck you, and then we fuck your women! *Allahu akbar! Allahu* ...'

The sound commander drowned him out. 'Complete load of bollocks!' I started yelling, but Sticky's Brother grabbed the microphone.

'You meant to say that you *fight* like women!' he yelled. 'In fact the British soldiers' wives fight better than you can!'

Sticky's Bro was grinning from ear to ear. I was praying for the hidden enemy fighters to show themselves. I kept checking my Rover terminal for any sign of their presence. As soon as they were spotted I'd get the call from the air, and we'd mallet them. Stacked above I had two Apaches, a solo A-10 Warthog, a pair of Harrier jets, and a B-1B up high. Each aircraft was scouring the ground, and

I was just waiting to get the call. We were the priority air mission for the night, and I had reserve air until morning.

'Hadin's a British spy, you daft clefts!' one of the 2 MERCIAN lads yelled out. Everyone was getting into it now. 'Get wise! Hadin's on our bloody payroll.'

Suddenly, the radio chatter went wild, as Commander Hadin himself responded: 'Don't listen to them, brothers! It's all propaganda and lies! Don't listen!'

The OC ordered the platoon to open up from the rooftop with the 50-cal, and the Gimpys. Blasts of tracer went arcing into the night, sending fingers of hungry fire groping towards the hidden enemy positions. When there was no response, I got Sticky's Brother to yell out the Pashto equivalent of: 'You're a bunch of fucking fannies: you won't even fire back at us when we're itching for a fight.'

Again the enemy chatter went mad. 'They're trying to provoke you to open fire!' Commander Hadin was yelling. 'Don't fire! Hold your fire! They'll see you if you open fire!'

'Hadin told us you'd come out to fight tonight!' Sticky's Bro yelled. 'But looks like you're soft as shit! What are you, a bunch of men or a bunch of girls?!'

I gave Sticky's Brother an approving grin. He gave me a cheeky smile in return. All we needed was one round to be fired, and the eyes in the sky would detect the muzzle flash, and I could call in the bombs.

'Hold firm! Hold firm!' Hadin urged his men. 'Stay in your positions! Hold your fire!'

At this point the Apaches picked up two heat sources moving through the trees. I glanced at the fluorescent blue-green glow of my Rover screen. Sure enough, two fuzzy heat blobs were creeping towards us, but no weapons were visible. No Afghan civvies would be creeping through Golf Bravo Nine One at night, of that I was certain, but under the rules of engagement we couldn't just take them out. I told the Apache pilots to fire warning shots, then fly off into the open desert as if they were leaving.

Moments later the image on my screen erupted in a shower of white-hot sparks, as the 30mm cannon rounds tore into the earth to one side of the target. The heat blobs froze, and a second later they had disappeared. They'd either covered themselves in blankets, to hide their heat signatures, or snuck underground. They didn't have to get very deep before the infrared scanners on the aircraft would lose them.

Gradually, the sound of the Apache's rotor blades faded away on the hot night air. The jets were too high and too distant to be audible.

'We've even sent the helicopters away!' Sticky's Brother yelled, once the sky had fallen silent. 'Maybe now you'll be brave enough to fight us!'

Still there was no response. It was getting a bit frustrating. I suggested to Major Hill that we tell the enemy that we had them surrounded. They just might get a flap on and open fire. The OC thought it a grand idea. I was about to do just that, when Sticky's Brother reported a new item on the intercepts.

'Commander Hadin's saying that he has us surrounded. He's saying he's got Alpha Xray surrounded.'

For a moment I stared at Sticky's Bro in confusion. 'Hold on a minute, that was our idea ... We were supposed to say that.'

Then the first spine-chilling wolf howls rent the darkness. Scores of enemy fighters started calling to one another, and as they did so we realised they *were* all around us.

There's not a lot that's spookier than the way the enemy do these animal howls at night. The chorus went up from one fighter to another, and on and on and on. They had crept in unnoticed and they had us surrounded. They sounded close, like spitting-distance close.

I locked eyes with Sticky. 'Where the fuck did they come from?'

He shrugged, keeping one eye glued to his night scope. 'Maybe we should've played them *Barney the Dinosaur* after all.'

The spine-chilling howls went circling around and around the darkened rooftop. I felt the hair on the back of my neck go up. We'd set out to trap them in the open and waste them, but instead they'd encircled us.

'*Widow Seven Nine*, all call signs in my ROZ,' I spoke urgently into my radio. 'Can you see anything? They're all around us at Alpha Xray.'

'*Bone Eight One*, negative.'

'*Recoil Seven Four*, negative.'

'*Hog One Five*, negative.'

'*Arrow Six Seven*, negative.'

Arrow was the US Apache's call sign. Bloody fantastic. That was two Apaches, an A-10, a pair of Harriers and a B-1B – and none of them had seen a sausage. Somehow, the enemy had crept in right under our noses without being detected. So much for the sound commander ruse, and our eyes in the sky.

'Is that the best you bunch of women can manage?' Sticky's Bro bawled into the microphone. 'Instead of howling like dogs, why don't you come out and fight us?'

We carried on abusing them for a while, but all we got in answer were those wolf cries. If we couldn't beat 'em, maybe we should join them. I started howling into the microphone. Sticky's Brother stared at me for a second, and then he was laughing his wheels off.

The rest of the lads on the rooftop got right into it, and soon we were all howling away. So there we were – them barking at us and us barking at them. It truly was barking. After a while the enemy still hadn't opened fire, and we realised that our howls weren't cutting it. I reckoned we should've brought Woofer with us, and given him a go on the sound commander.

Woofer had stuck with us through thick and thin. No matter how many times PB Sandford got hammered by mortars, 107s or RPGs, Woofer never deserted us. The lads appreciate real loyalty, and we had warmed to Woofer as only British soldiers can. He was fat and sleek and healthy, in spite of all the running for cover from the bullets and bombs.

In fact, PB Sandford was getting to be a right zoo. A flock of ducks had colonised the pool of dirty water next to the well. The lads were forever going and feeding them bits and pieces from the ratpacks that they didn't much fancy. The ducks were as happy as pigs in shit, and they'd recently been joined by a bunch of noisy, nosey chickens.

Plus there were all the stray cats that the lads were feeding with their leftovers. I'm not such a cat man myself, but I did kind of warm to the chickens. But what won top prize had to be Throp's donkey. One day he'd turned up at the gates riding bareback on a donkey, a set of makeshift reins gripped in the one hand. Throp's a big lad, and as he trotted back and forth on the ridge line his toes could touch the ground. He had the lot of us killing ourselves. What made it all the funnier was that Throp never once so much as smiled, or lost his composure. You'd have thought he was riding the Grand National, the way he carried on.

But no matter what animal noises we tried down at Alpha Xray, not a thing seemed to do it: we just couldn't get a rise out of the enemy. It never got close to kicking off, and by 0300 I'd lost all my air. There was nothing for it: along with the rest of the lads I dossed down on the dirt at Alpha Xray and slept like a lamb.

With Woofer to keep fed, Sticky had one more mouth to cater for. He'd taken to boiling up job lots of meals-in-the-bag in the empty GPMG ammo tin. The day after our warped night's howling at the enemy, Sticky went to hoof the lid off the tin to check on his cordon bleu cooking. As he did so, the pressure blew up in his face.

The ammo tin let rip and Sticky was left soaking wet and howling. The lot of us couldn't help it – we were rolling around on the dirt with laughter. That was until we realised how badly he'd been hurt. His face was blistering up, as were his forearms. Sticky was so bad he had to be casevaced to Camp Bastion. Word was that he'd be out of action for a couple of days at least, and the medics were worried that he'd have permanent scars. Still, the show had to go on.

TWENTY FIVE
JASON'S MAD MISSION

At stand-to the following morning I had an A-10 check in to my ROZ. As per usual, the airwaves were going crazy, with Commander Hadin urging his men to 'ignore the jet, and prepare to attack the main camp of the enemy'.

We'd heard it all before, and most of the 2 MERCIAN lads went back to their pits to get some extra kip. But the intercepts were as good an excuse as any to give those lads a second early-morning wake-up call.

'*Hog Zero Three, Widow Seven Nine,*' I radioed the pilot above. 'Sitrep. We're TIC-imminent. I need a low-level show of force over the top of our position.'

'Yes, sir. Commencing my run-in now for a show of force.'

I stood on the roof searching all around me for the Warthog. I could hear the growing whine of its jet engines as it swooped in towards us. All of a sudden the squat, ugly black form of the jet reared out of the sunrise, wingtips just over the HESCO walls.

The pilot screamed over the compound, banking around the sangar, the tidal wave roar of his passing smashing into the base. From JTAC Central the pilot was level with me, he was that low.

I was just getting on my TACSAT to congratulate him on a first-class show of force, when the lads came tumbling out of their beds, fully tooled up and ready to rumble.

Everyone was screaming all at once: 'What the fuck's going on?' Then they noticed me on the roof doubled up.

'What the fuck?' someone started yelling.

'What're you up to, you wanker?' another cried.

'We're TIC-imminent,' I managed to choke out. 'I had to do a low-level …'

'Fuck off we are!' one of the lads cried.

'All the fucking terps are asleep!' roared another.

'You crazy fucker!' yelled a third.

'We are, lads,' I insisted. 'We're TIC-imminent. It was all for your own good.'

Later that morning some elders turned up at the gates. The top news was that we'd smashed a lot of enemy with the air missions over the last few days. It was amazing how much Intel we could get off the locals. They wanted the Taliban out of the Triangle just as much as we did, and would risk brutal reprisals against them or their families. More often than not we'd be wandering about in the Green Zone, stop to have a chat with someone, and there'd be a peachy bit of information passed over.

Whenever we were out on patrol and it wasn't kicking off, the kids would gather round. The local nippers were fascinated by us lot of pasty-faced foreigners carrying all this space-age kit. We'd hand out the sweets, and as we did so I'd try and whip the caps off the nearest boys' heads. The locals wore these skull caps beaded in all different colours. A lot of the boys looked to be around seven or eight years old, the same kind of age as my Harry. Like him, they were cheeky and curious, and they loved playing that dodge-the-crazy-foreigner-hat-grabber game.

It was crap for security, of course, for any one of those kids could have been a suicide bomber. But it was great for hearts and minds. It was obvious you couldn't win over the locals without having contact with them, and there was no better point of contact than the kids.

The locals wanted the Taliban out of The Triangle just as much as we did. And once they were gone, the elders wanted things to go back 'to normal', which meant they wanted pretty much the same as we did. With things 'back to normal' we'd all return to a life at home with our folks. We'd be reunited with wives and kids and girlfriends, and put Helmand behind us. And here in the Triangle the locals could plant their crops, feed their families and put their kids through school, in the hope that life might improve in the future.

But not with the Taliban holding sway. Those bearded lunatics wanted to force the country back into the Dark Ages. In some warped, pseudo-medieval throwback, girls would be banned from school, women treated like cattle, and non-Muslims declared the eternal enemy. Under their rule, the most a young lad like my Harry could aspire to was to blow himself up, with the promise of seventy-two virgins to follow.

With the elders warning us that Commander Hadin was shipping in new fighters, the fight looked to be far from over. Sure enough, that evening Alpha Xray got whacked. It was 1900 hours when a volley of RPGs smashed into the base walls. With each new hit the cracks in the rooftop were getting ever wider, and new splits in the walls kept appearing. Sooner or later one of those rocket-propelled grenades was going to smash open AX, and cause carnage in there.

My fellow JTAC across the river at PB Arnhem had a pair of Apaches on hand. I borrowed one, and got it fast into the overhead. Chris ordered the mortar and artillery units to go to 'check fire' – to cease firing – so the gunship could hunt out the enemy.

As soon as the Ugly call sign was audible, the firefight died to nothing. I got the Apache over Golf Bravo Nine One, and within minutes the pilot had detected four enemy fighters. The Taliban were nothing if not creatures of habit: the stupid sods had been hitting Alpha Xray from the crater left by a previous thousand-pound JDAM strike. I still had the coordinates of that airstrike in my little black JTAC log: I passed them up to the Apache, and cleared him in to do four strafes with 30mm. The enemy were hardly able to scramble up the sides of the crater, before all four of them were whacked.

Using a bomb crater as 'cover' was like a dog returning to its own vomit: bad idea. But the one thing I couldn't understand was how the enemy kept getting into position at Golf Bravo Nine One without being spotted. One moment it was deserted, the next they were there, pounding Alpha Xray. Somewhere, there had to be a tunnel or a hideout.

The following morning I got allocated a Predator for six hours solid. It wasn't my favourite platform, but I'd not forgotten the Hellfire Thirteen strike, and I was determined to man it out. I got it over Bin Laden's Summerhouse, and once again there were scores of males of fighting age all around it. Frustratingly, not one of them had a visible weapon.

Next, I got it flying air recces over the Golf Bravos. I'd searched all the way down from Golf Bravo Nine Eight to Golf Bravo Nine One, and was just moving on to Alpha Xray. I wanted to give the lads there some stick about my being able to see them without their helmets or body armour on – just to let them know that they weren't forgotten.

But as the Predator cruised south-west I noticed something odd. Adjacent to the bald brown scoop of the thousand-pound crater was a tiny thread of smoke. I wouldn't even have seen it had I not been studying the crater so closely. It looked like a cooking fire. But why on earth would anyone be getting a brew on in the middle of a blasted battlefield?

The only person likely to do that had to be an enemy fighter. I got the Predator to zoom in on the thin column of smoke. At its base was a tiny, two-metre by four-metre mud-walled building. It had been totally covered by trees, until the JDAM had blown away enough foliage to partly reveal it. This had to be the entrance to the enemy's hideout, from where they kept popping up to hit us. I felt certain of it. I just needed an excuse to smash it from the air. I got the Predator to pass me the ten-figure grid of the tiny, bunker-like building, and stored it away for later.

Once I'd lost the Predator I got *Rammit Six One*, a Dutch F-16, into my ROZ. I passed the pilot the ten-figure grid of the bunker, and asked him to zoom in his sniper optics.

'*Rammit Six One*, tell me what you see,' I radioed.

'Roger. Searching.' There was a minute's silence, then this: 'Visual a tiny building, but only when banked off to the east. Obscured by foliage otherwise. There's a small fire beside it.'

So now I knew we had their hideout nailed. All I had to do was pass an aircraft those coordinates, and they'd be on to it.

At 2030, just after last light, Alpha Xray got smashed again. Machine-gun fire was whipping out of Golf Bravo Nine One, plus an RPG team were hitting the base from a new position to the north.

I got allocated a pair of Mirages, my least favourite platform. Top joy. The French pilots weren't familiar with the area, so I gave them a full update, then requested: '*Rage Three Two*, I want a 500-pounder dropped on that RPG team to the north of Golf Bravo Nine One. Nearest friendlies to the south-west one-two-five metres.'

'Negative,' the pilot replied. 'It is too close and I cannot drop ordnance.'

'Shut up,' I snapped. 'We'll just use you on a north–south attacking run, which'll keep the blast away from friendlies.'

'Negative. I 'ave never dropped this close to friendlies. It is beyond danger-close and in darkness, and ...'

'Break! Break!' The pilot of the other Mirage cut in. '*Widow Seven Nine*, *Rage Three Three*. Visual twelve pax to the east of Golf Bravo Nine One, all with weapons, moving south towards Alpha Xray. Now going firm: grid is 04827436.'

I confirmed the grid and checked my map. Those twelve fighters were now the nearest and single greatest threat to AX, and I wanted them smashed.

'*Rage Three Three*, I want a 500-pound bomb dropped on that grid on a south-west to north-east attack run, to throw blast away from friendlies. Confirm.'

'Negative,' the pilot replied. 'I cannot do the drop.'

'Listen,' I rasped, 'every other platform in theatre has done drops this close and closer, so why the bloody hell won't the both of you?'

'It is just too close,' the pilot replied.

It was time to call their bluff. 'Rage call signs, you are not up to task. I'm calling for replacement air.'

'*Widow Seven Nine, Rage Three Three*. I will come in and do the drop, but first I need the ground commander's initials, and your name.'

'Ground commander's initials SH. I'm *Widow Seven Nine*.'

'No, no – I need your name, please,' the pilot repeated.

'I just told you, I'm *Widow Seven Nine*.'

'No, I 'ave to 'ave your real name please …'

By the time the French pilot had finished buggering about, half the heat sources had disappeared. I talked both jets in, got them to do four drops and winchestered them. It was six minutes after we'd started arguing, and they were all out of bombs.

'BDA: we have killed nine enemy pax with the four GBU-38s,' the pilot of *Rage Three Two* reported, sniffily.

Frankly, I didn't believe a word the Rage pilots were saying. I just wanted them gone and some different air above me. I got on to Widow TOC.

'Both Rage call signs are fucking Winchester. The contact's still hot, so now can I have some proper air?'

The Mirages were ripped by *Rammit Seven Three* and *Rammit Seven Four*, a pair of Dutch F-16s. We hit a score of enemy positions danger-close and without any problems, and finally the contact died down to nothing. I went to my cot cursing the French pilots, and wondering how on earth we were going to smash that enemy bunker at Golf Bravo Nine One.

At breakfast the following morning, Chris, Sergeant Major 'Peachy' Peach and I formulated a plan. I had air allocated for later in the day. I'd get it banked off to the desert in the south, where the enemy couldn't hear it. I'd pass the pilot the ten-figure grid of the bunker, and Peachy would get in a WMIK and start driving. He'd head down Route Crow, pass Alpha Xray and keep going. Pushing onwards he'd hit the southern boundary of Golf Bravo Nine One. If his arrival didn't provoke the enemy to open fire, he'd start malleting the bunker with the WMIK's 50-cal. Just as soon as they opened up he'd reverse like fuck, and I'd get the air to smash the bunker.

It was proper British tactics, and I had no doubt that the sergeant major would do it. He had bollocks the size of a horse. It was just his kind of thing, and reminiscent of the crazed mission he'd driven in the WMIK, when he rescued the three wounded men during the battle for Rahim Kalay.

When the three of us put our plan for smashing the bunker to the OC, he just stared at us for a good long second.

'I just don't want to fucking know,' he said.

We guessed we'd got the go-ahead. At 1300 I had a pair of Dutch F-16s – *Rammit Six One* and *Rammit Six Two* – check in to my ROZ. I passed them the grid of the bunker. The Rammit call signs seemed happy enough with the plan of attack, but the pilots did query what would happen if the WMIK got bogged in or rolled.

'If that happens we'll abort the drop,' I told them. 'I'll get you to do a low-level pass, as the lads make a run for it.'

'Roger. Banking to the south now.'

I gave Peachy the thumbs-up, and he set off in one of the WMIKs down Route Crow. He had a volunteer at the wheel, and was himself manning the 50-cal in the rear. I got the F-16s to stand off six nautical miles away, so around a minute out from target. I was up on the roof with eyes on, and Peachy was down around Alpha Xray.

'You ready?' Peachy queried over the radio.

'Aye,' I replied.

'Right, I'm off,' said Peachy.

And that was it – he set off like the charge of the Light Brigade. From JTAC Central I could see the WMIK bucking and kangaroo-ing over ruts and ditches, as it careered towards Golf Bravo Nine One. Fifty metres short of the target the bunker just seemed to erupt, as fighters swarmed out and opened up on the WMIK.

'FUCKING CONTACT!' Peachy screamed into his radio. 'ENGAGE! ENGAGE 'EM!'

As the WMIK slammed to a halt and began a crazed retreat in reverse, Peachy was crouched over the 50-cal malleting the bunker.

I dialled up the F-16. '*Rammit Six One* – I need you to hit that bunker now!'

'What about your vehicle? Is it a safe distance?' the pilot queried.

'Don't fucking worry about that, worry about the drop!' I yelled.

'Tipping in. Thirty seconds,' the pilot warned me.

'Thirty-second warning!' I screamed at Sticky. Then at the pilot: 'Friendlies sixty metres west of target – those lunatics in that WMIK. You're clear hot.'

'In hot.' A beat. 'Stores.'

The GBU-38 five-hundred-pounder was on its way. It was now a race against time as Peachy's driver gunned the WMIK, and the bomb came howling in. Just as the vehicle careered behind the HESCO wall at Alpha Xray, the JDAM slammed into the roof of the enemy bunker.

There was a massive, blinding flash as it detonated, and an instant later the treeline erupted in a fountain of smashed walling, splintered tree branches and flying dust and shrapnel. The blast wave tore across the roof at PB Sandford, and as the smoke cleared at the target we were all eyes on the point where the bomb had hit. There was nothing left of the enemy bunker but a massive smoking hole. It had been completely obliterated. I didn't need a BDA, but I did pass up a heartfelt well done to the Dutch pilots. With that Peachy drove back to PB Sandford.

'Job done!' I yelled at him, as soon as he was back with us. 'Done 'n' dusted. A fucking beauty, mate.'

'They must be fucking wounded,' Peachy grinned. 'One moment they see the jets disappear, then there's this lunatic suicide driver coming towards them, and they think – *Fuck me, this is it!* The next, out of nowhere they get splatted!'

Peachy, Chris, Sticky, Throp and I were just congratulating each other on a job well done, when Sticky's Brother came dashing over to us. He was so excited that he could barely get the words out.

'New item on the intercepts,' he announced. 'Everyone keeps asking for Commander Hadin to check in.' Sticky paused for dramatic effect. 'No one is answering. *No one!*'

Peachy and I locked eyes. Fucking hell. I could see why Sticky's Bro was excited. It looked as if we'd just smashed the top enemy commander in the Triangle, the guy who'd replaced Jamil.

And if we had, then Jason's mad mission had been a proper peachy one.

TWENTY SIX
BAD NEWS FROM PIZZA PIE WOOD

From that moment on the attacks from Golf Bravo Nine One ceased completely. There was a deathly hush over the Triangle. I guess the enemy had to be licking their wounds.

Sticky's Bro had heard Hadin speaking on the radio, so Peachy's mad mission hadn't killed him outright. But the Intel was that he was wounded, so we were halfway there. Plus the veteran Taliban fighters were twisting on about getting replaced. Like us, the enemy had a system of posting fighters for an allotted time, and then relieving them. I'm not too sure where they went for their R&R: probably the alcohol-free seventy-two Virgins Theme Park, across the border in Pakistan.

It was a couple of days after Peachy's mission when things started to get busy again. It was an oven-hot boiler of an afternoon, when one of the lads spotted an Apache coming out of the haze to the east of the valley. I got on the TACSAT and dialled up the pilot. We had a foot patrol halfway down to Alpha Xray, and if nothing else I wanted to make sure he didn't mistake them for the enemy, and blast 'em.

It was two American Apaches – call signs *Arrow Two Three* and *Arrow Two Five* – and they'd just been flying air recces over PB Arnhem. They asked me what I'd been up to, 'cause word was getting around that my call sign, *Widow Seven Nine*, had been busy. Then they offered me thirty minutes' playtime.

I got the pair of Apaches flying recces over the silent valley. For twenty minutes nothing was seen, so I asked the pilots if they'd do a low-level fly pass at PB Sandford. We wanted a cheesy photo of the FST with the gunships in the background. The American pilots seemed more than happy to oblige.

We gathered the FST together, but left Jess sleeping in his bunker. We reckoned it'd be a great wind-up to deliberately leave Jess out of the FST photo. We got tooled up in our full battle rattle and clambered up to JTAC Central. We asked Paddy – an Irish lad who'd been helping out with the FST – to take the shot.

I talked the pilots through what I wanted.

'We're after a good souvenir photo, with the four of us in line and you guys coming in from behind. Come in real low but to the west, to keep you out of the sun.'

'No dramas,' the pilot confirmed. 'Banking around now.'

As the Apaches thundered low across the bush, there was a sharp crackle of gunfire. Bullets went tearing past our heads. Rounds started kicking up the dust and dirt on the roof. We forced a line of cheesy grins, and yelled at Paddy to take the shot before someone got their head blown off. As soon as he was done we broke ranks and dived for what little cover there was. The lead Apache was passing over the Golf Bravos, and from below it a stream of fire tore upwards at the gunship. One half of my mind was thinking: *Bollocks, that's the end of our photo op.* The other half was thinking that I'd better warn the aircrew.

'*Arrow Two Three, Widow Seven Nine*: you've just been engaged by small arms fire from just to the south of Golf Bravo Nine Two.'

'Roger. Stand by.'

I cleared it with the OC to get the Apaches to fire warning shots, and radioed the pilot.

'*Arrow Two Three, Widow Seven Nine*: I want you to fire warning shots at the position you were engaged from.'

'No, sir,' came the reply. 'I am conducting my reconnaissance.'

Fair enough. The pilot was obviously having a good look, and scanning for targets.

There was a sleepy call from below us: *What's going on?* It was Jess, fresh awake from his cot with all the gunfire. As he clambered up the steps he realised the entire FST bar him were on the roof all tooled up with guns, body armour and helmets.

'What's going on?' he repeated.

'Erm … We wanted a photo op with the Apaches …'

'You were asleep, so …'

'Didn't want to wake you, mate.'

'Reckoned you needed your beauty sleep.'

Jess looked totally devastated. 'You did the FST photo without me?'

I was starting to feel a bit bad about the wind-up. There was a fresh burst of shooting from below. The second Apache was now taking fire. I radioed the pilot and told him what was what, and asked him to put down some warning shots.

'No, sir,' he replied. 'I'm conducting my recces.'

There was a call from the OC. 'Bommer, what's going on with those Apaches?'

'Sir, they say they're conducting …'

Thump-thump-thump-thump … My words were lost in the pounding percussions of 30mm cannon fire. I turned to see *Arrow Two Three* spitting flames from its chin turret, and churning out a thunderous burst of rounds. At the same time there was a savage stab of fire beneath *Arrow Two Five*'s stub-wings, and two Hellfires were away.

Fuck me. The Arrows were opening up with all they'd got, and we still had a foot patrol out in the bush. Not only that, but neither of the bastard pilots had cleared it with me.

The US Apache pilots worked on different rules of engagement to our own, which gave them the right to open fire if they believed themselves to be under threat. They were taking fire, and they'd opted to return fire – but it would have been nice if they'd told me what their target was.

'Bommer!' the OC was yelling. 'What the hell's going on with those Apache?'

'How the hell do I know?' I yelled back. 'I didn't ask them to fire a bloody thing!'

'Arrow call signs, *Widow Seven Nine*!' I was yelling. 'Arrow call signs, what the fuck are you two firing at?'

'Wait out,' came the reply.

The pair of gunships were now no more than three hundred metres above us at PB Sandford. They were going crazy firing cannons and Hellfires into the valley below. I kept calling for a sitrep, but all I ever got was a *wait out*. Spent 30mm casings rained down on us, as the pair of gunships let rip in one long, uninterrupted shooter-shooter burst.

The lads were staring at me, as if it was my fault we had two lunatic Apache pilots shooting up the Green Zone. Within five minutes both gunships had winchestered their 30mm cannons, and only had two Hellfire left between them. I'd counted *Arrow Two Three* doing eleven runs with its cannon, and firing four Hellfire, and the other gunship was going in equally hard. I'd never in my life seen anything like it. Everyone kept yelling at me: *What the fuck are the Arrows up to?* I still didn't have a bloody clue what they were shooting at, as the pilots weren't answering me. Finally, the pair of gunships ceased firing.

'Listen, Arrows, you pair of …' I yelled into my TACSAT. 'It's my ROZ and you need to get bastard clearance! We've got a foot patrol out on the ground …'

'Sorry, sir,' the pilot cut in, 'but we had thirteen armed pax in the treeline and we were flanking them so they couldn't escape.'

'Say again?' I asked.

'We've just killed thirteen males of fighting age,' the pilot repeated. 'We've killed thirteen minimum, and we need to bug out, 'cause we're sippin' on air up here.'

'Well, cheers, but next time we work together let me know what you're fucking doing!'

'Affirmative, sir.'

And that was it; the Apaches were off.

I turned to the OC. 'That was the Arrow pilots. They've just killed minimum thirteen enemy in the treelines south of Golf Bravo Nine Two.'

'*They've what?*' said the OC.

I gave him a look, as if to say: *I know, the bloody lunatics, but it was nowt to do with me.*

'*Widow TOC, Widow Seven Nine,*' I spoke into my TACSAT. I was wondering how they were going to react to this one. 'Sitrep: *Arrow Two Three* and *Arrow Two Five* are leaving my ROZ, winchestered. They've conducted twenty-one strafing runs and fired six Hellfire.'

'*Widow Seven Nine, Widow TOC:* say again.'

I repeated the message, and then Damo Martin came up on the air from FOB Price.

'Shut up, Bommer man, you're just being a dick. What's the score with the Apaches?'

'Mate, both gunships have just left me fucking winchestered.'

There was a moment's silence, then: 'Well what the fuck have you been shooting at?'

'According to the pilots we have minimum thirteen dead in the trees to the south of Golf Bravo Nine Two.'

Neither Damo nor Widow TOC would believe me. Instead, they decided to send a Predator over to check. I was told I'd have *Overlord Nine Five* above me in four minutes. By now, Sticky's Brother was also getting noticeably agitated.

'They're calling for Commander Hadin to check in!' he shouted. 'Over and over and over. Commander Hadin! Commander Hadin! Commander Hadin!'

His enthusiasm was infectious. 'Let me guess – Commander Hadin's not answering?'

Sticky's Bro nodded, gleefully. 'And there are lots of other commanders they keep calling for, and they don't answer either.'

No one seemed willing to believe me about the Arrows – not until the Predator operator got it over the attack site. There it found eight bodies lying beside the treeline, all with weapons. The analysts reckoned there were at least five more corpses half hidden in the trees. So that pretty much confirmed what the Apache pilots had said.

A while later I got another call from Damo Martin. Intel had come down from on high with the Americans. The thirteen kills were confirmed, and amongst their number were six enemy leaders – Commander Hadin included. The Taliban top brass had been in the midst of doing a handover with their troops, when the Arrows had hit them.

'Top bloody job, Bommer, mate,' Damo kept telling me. 'Top bloody job.'

'Mate, I didn't do owt,' I tried to object. 'I did not call a single shot. The Arrows just went lunatic and malleted everything.'

'Shut up,' Damo countered. 'You're just being bloody modest.'

'Mate, I did not call one single shot.'

'Shut up, you tit! I'm having none of it.'

Whatever I tried to say, Damo wouldn't believe me.

It still wasn't a top fluffy feeling for me though. Alpha Xray was only a hundred and twenty metres from the attack site, we'd had a foot patrol out on the ground and I'd had no idea what those Apaches were up to. But all was well that ends well.

Later, I'd settled myself down with a Flashman book that my Light Dragoons mate, Spunky, had lent me. I was reading by the light of my head torch. It was just after midnight and I was enjoying the warm afterglow of those Arrow airstrikes. Flashman was a Light Dragoons man, of course, and I couldn't help but love the bloke. I'd grown a pair of 'lamb chop' whiskers, just like him, and it was his tales of derring-do and caddishness that were helping get me through Afghanistan.

There was a call on the TACSAT. It was a covert surveillance aircraft, *Bat Zero Two*, and I had him for two hours above me. I wasn't very happy at being torn away from Flashman's adventures, but I heaved myself out of my cot and into the Vector.

The aircraft picked up a new piece of comms: *The heavy weapon is here and ready for pick-up in the desert*. I got Sticky to waken Chris who woke the OC. We reported it up the chain, and I got allocated two A-10s – *Hog One Two* and *Hog One Three*. The A-10s would be with me in thirty minutes. Meanwhile, *Bat Zero Two* was picking up all sorts of stuff about the heavy weapon handover. With the Warthogs still twenty minutes out, I feared we'd miss the bloody thing.

As soon as they were in my ROZ I got the A-10s flying recces north-east, scouring the desert. A heavy weapons handover meant vehicles, and that's what we were looking for. As the Hogs searched all up the desert with their sniper optics on wide-field view I had my eyes glued to my downlink.

But it wasn't long before I knew for sure that we'd missed them. Sometimes, you just sense these things. Then we had it confirmed on the intercepts: *The heavy weapon has been successfully delivered*. I lost the A-10s and the surveillance platform, it was 0245 and I was alone with Flashman once more. I wondered what it was that they'd sneaked into the Triangle to hit us with. It would not be long before I found out.

After stand-to the OC gathered the lads for a briefing. He told us that the rules of engagement had just been changed, which meant we could only fire on the enemy when they were firing at us.

We were holding the front line in the Green Zone, but we were forbidden from engaging armed enemy fighters unless they were actually in the process of firing at us. In practice this meant the enemy could gather for an assault on one of our bases, and we couldn't hit them until they opened up on us. It was like fighting with one arm tied behind your back. It felt like a right kick in the knackers, but we tried to shrug it off. The one consolation was that we were nearing the end of our tour: ten more days of the Triangle and we'd be out of there.

After months under siege, the strain was more than starting to show. We all of us had that wide-eyed, mad, glazed stare – the look that comes from day after day of adrenaline-pumping combat. Plus the lack of sleep was really starting to nail us. I guess I was particularly badly hit, as I was always getting allocated air in the middle of the night.

Barely a day had gone by when the enemy hadn't whacked the base with something – either 81mm mortars, 107mm rockets, RPGs or rounds. It was all mud here at PB Sandford: mud walls, mud floor, mud roofs and mud-filled HESCO barriers. Most of the incoming had smashed apart the mud a little more, but so far we'd been insanely lucky and no one had got splatted. Not yet, anyway. I guess the enemy were getting a bit frustrated at not killing us, and that was why they'd shipped in the mother of all weapons.

It hit us first at 1000 hours, in the midst of a big cricket-off. There was this distant, muffled bang, and then Mikey Wallis was screaming like a mad thing from his bunker.

'FUCKING INCOMING!' He hit the air horn: 'BWAAAAAAAAAARP!'

From Mickey's tone of voice I guessed this was something different. By now Throp and I had moved into a sandbagged bunker, which we used as our 'bedroom'. As we dived into the darkness, the howl of the incoming was like a bloody great big spaceship coming down to land on top of us.

Whatever it was slammed into the dirt fifteen metres to the front of PB Sandford. A massive explosion tore across the base. Even down in the bunker it was deafening. I felt the punching wrench of the shockwave tearing over us, the air being ripped out of my lungs, and then this thick cloud of smoke and dust came billowing down the stairway.

I forced myself to do the opposite of what instinct told me, and legged it for the Vector. I could barely see where I was going, but I fought through the burning smog. My TACSAT and all my kit was in the wagon, and I needed to dial up air. Whatever it was that had hit us Mikey would have the grid, and now was the time to find and smash it.

As I dialled up CAS, I could hear the crump of our own mortars firing, as they sent a counter-barrage on to the enemy's grid. Plus Chris was on the radio, dialling up a barrage from the 105mm howitzers. With the guns thundering away, I was told I had two Harriers – *Recoil Four One* and *Recoil Four Two* – inbound ten minutes.

The guns and mortars roared and snorted for five minutes straight, pounding the enemy grid, and then they ceased fire. Thirty seconds later Mikey was yelling again, and thumping his air horn.

'BWAAAAAAAAAAARP!'

A second monster projectile was inbound. No one had a clue what it was yet, but it sounded like a bloody great thousand-pound JDAM. Throp and I legged it for the bunker, as the incoming screamed through the burning blue right on top of us. It tore across the base, smashing into the desert twenty metres beyond the back wall.

Throp and I locked eyes. 'Fuck,' he said. 'Now they've got us bracketed.'

That was one warhead just outside the front wall, and one just over the back. They'd split the difference with the next, and it'd be bang on. We dashed for the Vector, and I dialled up the Harriers. I gave them the grid, and told them to get over it pronto looking for some kind of massive gun or rocket launcher.

The Harrier pilots got overhead the grid but there was nothing moving. There was a pizza slice-shaped stretch of woodland, alongside which was a thin treeline, which corresponded to the grid. I got the Harriers searching up and down that woodstrip, but not a thing was to be found. Where the fuck had they hidden that weapon?

For forty-five minutes the Harries scoured the grid, but nothing. I was forced to close the TIC and lose the air. A few minutes later Mikey let out a yell, and hit the air horn. We knew then that we had a third projectile inbound, and that we were bang in the centre of its path. As we sprinted for cover, I saw one of the radio operators come haring out of the radio shack with his canvas chair stuck to his sweaty arse. Another of the lads came pelting out of the shitters, with his trousers down around his ankles. It would have been funny, were this monster weapon not so terrifying.

The killer warhead snarled out of the empty sky and slammed into the mortar position at the rear of the base. It erupted in a whirlwind of shock and pain, smashing great chunks out of the HESCO and flattening a steel fence, before the last of the blast ripped into the tented medical centre. It was pure luck that no one was in there and that the mortar team had made it into cover. There would have been nothing left of anyone under that blast. A second round was fired directly after the first, and this one hit fifteen metres from the Vector, tearing the shower block to shreds.

Luckily, the projectile buried itself in the sand, which kept the frag down. But it was as if a thumping great earthquake was tearing the base apart. Even from the bunker, Throp and I heard the jagged chunks of steel whistling through the air. The trajectory of the thing meant it must have missed the Vector's roof by inches.

Throp and I came out of the bunker giggling crazily – but it was more from fear than good spirits. Fuck, that heavy weapon was horrible. And whatever it was, they had it zeroed in on us now.

If I couldn't nail it from the air, then it was going to tear us to pieces.

TWENTY SEVEN
199 KILLS

Chris and the OC went for a walk and a quiet chat about what to do about that heavy weapon. They came back with a jagged chunk of metal the size of a dinner plate, which they'd found in the mortar compound. From that they ID'd the weapon: it was a gigantic 120mm mortar.

The lads from our mortar team handed around that chunk of shrapnel, staring at it weirdly. They just kept shaking their heads, and going *fucking hell*. It was then that I realised what a truly heavy piece of shit it was. The looks on their faces said it all. Everyone was shit scared, and with good reason.

A 120mm mortar is a brute of a thing. The biggest mortar the British Army uses is the 81mm, and most of the lads had never even seen something as big as a 120mm. It fires a round the same calibre as that of a Challenger II main battle tank. It was like having one of those sat outside the gates of PB Sandford and tearing the base apart.

Normally, a 120mm mortar comes mounted on a chassis with car-like wheels, so it can be towed into battle behind a truck. The barrel itself stands taller than the operator, and it can lob its fin-stabilised rounds over seven kilometres. How the enemy were managing to fire off those monster mortars whilst hiding the launcher from the air was mind-boggling.

Being under that 120mm was hell. Whenever the banshee howl screamed down on us, it was like a lottery with death. It drilled into our heads, everyone running like mad for a bunker. But whatever cover you found, if a 120mm landed on your roof then you were a dead man. All that would be left of you was a shredded, bloodied pulp: that's if your mates could find anything. And the forty-five seconds the round took to arc through the air, all the while howling like a ghost train, felt like a bloody lifetime.

One of those giant shells landed next to the bunker that Throp and I were sharing. Luckily, neither of us was in there at the time. Jagged chunks of shrapnel ploughed through the roof and slammed into the dirt floor, tearing our mozzie nets to shreds, embedding themselves in the wall where I had a few photos of the family pinned up. *Well, that was it*. It was *personal* with that enemy mortar crew now. It was time to even up the score a little. Whatever it took I was going to fix them. I was going to nail those bastards. From then on each time the air horn went off I'd sprint in the opposite direction of any proper cover, and dive into the Vector, closely followed by Mikey Wallis.

The first time we tried to hold the wagon's hatches shut, as the shell screamed down on us and we mumbled our prayers. Trouble was I couldn't shut the one hatch, for the TACSAT antenna cable was sticking out of it. For a second I glared at it.

Then I pointed and said to Mikey: 'What the fuck do we do if it comes through that?'

We started laughing, hysterically. But it didn't sound very clever as the shell came howling in on us. We stopped smiling, and ka-fucking-boom, it smashed into the compound, blasted sand and shit slamming against the metal skin of the Vector.

Mikey had a fix on the mortar launch point. His ten-figure grid was 3.8 kilometres to the north-east of our position. I got allocated a Predator, and I got it to do an overwatch of the mortar firing point, with a live feed to my Rover terminal. After each 120mm round went up I saw half a dozen males of fighting age leaping over a wall and diving into a tiny building. Were they the mortar crew –

the bastards I so badly wanted to get? I was pretty certain they were, but I couldn't actually see the mortar firing, to 'positively ID' it. It must have been doing so through the shadow of a roof, or a door, or a window, but they had it too well hidden. I began wondering if they had some kind of automatic, rollable roof, so that as soon as it fired it slid it back into place again. This was more James Bond than Mullah Omer's Taliban. It was messing with my head.

That evening John Hill, Jase Peach, the OC, Chris and I were talking at the back of Vector. We were saying how fucking horrific that 120mm mortar was, and how we needed to come up with a plan to smash it. Everyone was shit scared of it: when the air horn went off it was the worst feeling possible.

We reckoned they wouldn't risk firing the thing at night, for then our air could track down the hot tube with infrared scanners. The OC decided to push a fighting patrol out towards the Golf Charlies, on foot and at night. His aim was to show the enemy that we weren't cowed by their mortar. But there was also the hope they might be tempted to lob a couple of 120mm rounds at us, in which case we could nail the hot tube.

The patrol left PB Sandford at 1900, heading across the high ground to Monkey One Echo. As soon as it was out the intercepts started going wild about the 'Diamond Special Forces' being out on foot in the Green Zone.

I had *Hog One Five* and *Hog One Six* in the overhead, and for this patrol we'd been granted less restrictive rules of engagement. By 2000 hours the patrol was pushing into the dense bush around Golf Charlie One Seven, and heading in the direction of Bin Laden's Summerhouse. At this point the lead edge of the platoon spotted three armed figures fifty metres ahead, in ambush positions. I passed the grid to *Hog One Five*, and told him to hit them. I asked him to attack with his 30mm cannon, on a north-west to south-east run.

'I want the strafe of all strafes,' I told him, 'all along that treeline.'

'Affirm,' he replied. 'Banking around.'

For thirty seconds or so you could hear a pin drop in the stillness of the night, and then the A-10 came screaming in like a thing possessed. When the pilot finally unleashed his seven-barrel Gatling gun I thought the strafe would never end.

'Brrrrrrrrrrrzzzzzzzzzzzzzzzzzzzzzzzzzzzzzzzzzzzzzzzzzzzzzzzzzzzttttttt ttttttttttt.'

It was the longest I'd ever heard, the 30mm thundering on and on as it ripped apart the treeline. The BDA was bang on target: two enemy dead, and a third dragging an injured fighter away. I told the A-10 that it was a class strafe, and that the platoon would pull back to Monkey One Echo, as we'd found the enemy's front line.

At that moment, I got a call sign trying to break into my radio traffic.

'Break! Break! *Widow Seven Nine*, *Widow TOC*: on no account are you to engage the enemy with the Hog call signs.'

'Say again,' I replied.

The message was repeated.

'Roger: why not?' I asked.

'Your rules of engagement you can only use with British jets.'

'Well, there's a couple of Taliban in the Green Zone'll probably wish you'd sent that message a few seconds earlier.'

'Why?'

''Cause I just killed two, and the third is dragging one of the injured away.'

'Stand by.' There were a few moments' silence, then Widow TOC was back on the air. '*Widow Seven Nine*, there might be a problem with that.'

'Not for fucking me there's not,' I told him. I flipped back to the A-10's frequency. '*Hog One Five*, seems there's something wrong with the engagement. I wasn't meant to fire 'cause of the rules.'

'What the … why?'

'Look, it's nowt to do with you guys. I bought the rounds, so if anyone's in the shit it's me. Can you watch over the patrol, while I try and sort this shit out.'

'Roger that.'

I flipped frequencies back to Widow TOC. 'Look, 95 per cent of all controls in Helmand are with US platforms. If what you've said is right, you should've made sure we had Harriers over the patrol.'

The duty guy at Widow TOC ducked the issue. '*Widow Seven Nine*, release the Hog call signs once your patrol is back in base.'

I told the Hogs I'd been ordered to release them, but they basically refused to go.

'We're on X-CAS, *Widow Seven Nine*, and we're remaining on X-CAS. And it just so happens we'll be right in your overhead, and it just so happens you'll still be receiving our Rover downlink. You OK with that?'

'Fucking cheers, lads,' I told them.

I'd worked with these pilots before, and we knew the score. But I wasn't about to let it rest with Widow control. I was steaming.

'*Widow TOC, Widow Seven Nine*. Look, we need to clear this up. Are you asking us to hold the front line in the Green Zone, but if we find armed enemy we're not allowed to do anything? 'Cause if that's the case you need to get the lot of us out of here.'

'Stand by.'

As I waited for a proper answer, I got a call from the A-10s.

'Sir, are you watching your Rover? I'm visual with an eight-man patrol with RPGs and AKs moving west out of Qada Kalay.'

I flicked my eyes to my Rover screen: the eight-man enemy patrol was clearly visible, snaking through the trees.

'Roger. Stand by.' I called Widow TOC, told him what we could see, and asked if we were clear to engage.

'Negative: they are not an immediate threat to you.'

'Not now they're not,' I fumed, 'but what about when they reach Alpha Xray in an hour's time?'

'Negative: they are not an immediate threat to you.'

I got on to the A-10 pilot, and told him what was what. Then: 'Is there any chance you can dive on to target, have an ND with the 30mm, and mow the lot of them down?'

The pilot burst out laughing. An ND stood for negligent discharge – a posh term for firing off some rounds by accident.

We tracked that patrol for an hour or more, as they passed through several enemy checkpoints. Every five minutes I kept asking for clearance to fire. I didn't get it, and finally the A-10s were out of fuel.

'Stay safe,' the pilot told me. 'It's excellent work you're doing down there, *Widow Seven Nine*, we all know that. Sometimes the rules are just shit – what can you say.'

The A-10s left my ROZ and I lost the downlink. Whilst I'd been controlling the jets, Throp had been killing time by tallying up the kills in my JTAC log. He turned to me with a grin.

'Guess what?' he said.

'What?'

'You're not going to fucking believe this, Bommer, but you've got more kills than Harold Shipman. You're on 199, mate. *199*. We've got to get over two hundred.'

I'd never once thought about totting up the kills. I guess I'd been too busy doing them. In any case, body counts always have a degree of inaccuracy in them and frequently get overestimated. And there were better measures of our success – like the fact we'd seized and held the Triangle for many weeks now.

Throp pointed out that it was 199, not counting the thirteen killed by the crazed Arrow Apaches, plus the thirty-four MIAs reported by the elders after the battle for Rahim Kalay. He was determined that we'd top the official two hundred mark by the end of our tour. We'd have to get busy. We had barely a week left in the Triangle. But before we could kill any more of them, they were going to have a seriously good try at killing us.

At first light the 120mm was back in action, pounding PS Sandford with a murderous barrage. I knew it was only a matter of time before someone got killed. I'd been with the 2 MERCIAN lads for five months now. We were deep in Helmand, besieged and surrounded in the heart of bandit country, yet to a man the lads had taken it all in their stride. These young soldiers were Britain's finest. They were true warriors. And they were relying on me to nail that enemy mortar crew. I decided that there was only one way to kill that mortar team. I'd have to flatten the entire six-hundred-metre-square grid located by Mikey's radar gizmo. I dialled up a B-1B that was inbound into my ROZ. I'd had good dealings with the pilot before. We'd bonded over a couple of big actions. I reckoned he'd be up for what I had in mind.

'*Bone Three Seven*, this is *Widow Seven Nine*, d'you copy?'

'This is *Bone Three Seven*, nice to be working with you again, *Widow Seven Nine*. What can I do for you, sir?'

'I've got a 120mm mortar located to grid ref 1798617486. You'll find it in the Pizza Pie Wood area on your GeoCell map. It's zoomed in on our position, and we're getting murdered down here. You reckon you could flatten that entire grid?'

'No problem, sir. If that's what you're wanting, just give me the word.'

'What're you carrying?'

'I've got a full load, that's seventeen munitions. Two two-thousand-pound JDAMs, two one-thousand-pounders, and thirteen five-forty- and five-hundred-pounders. I can saturate the entire grid, if that's what you're after.'

'Aye, too right it is – your full load. You get clearance your end, I'll get clearance mine.'

'Roger that. Standin' by.'

I put a call through to the Widow TOC. I explained that we had the 120mm mortar located to a ten-figure grid, and that I had a B-1B on standby to flatten it. I asked for clearance to proceed. Unfortunately, the duty operator at Widow TOC didn't seem to understand what I was asking for, or why.

'What munitions exactly are you intending to hit them with, *Widow Seven Nine*?'

'The entire bloody lot, mate.'

'The entire ordnance package of a B-1B?'

'Aye. The Yank pilot's happy enough, so am I cleared or what?'

'But that's ... four-point-five million dollars' worth of bombs.'

'Listen, mate, I don't care how much it costs – am I cleared?'

'Negative. Not for one mortar, no. Why can't you just use one JDAM?'

'Listen, mate, I don't think you get it. Ever been pinned down by a 120mm mortar? Any idea what that's like? I need to flatten the entire grid, before one of us lot gets killed. That's what the Yank pilot has agreed to. So am I cleared, or what?'

'Negative, *Widow Seven Nine*, do not proceed with the attack.'

'You're saying I can't hit it, is that what you're saying?'

'Affirmative.'

'Listen, mate, get me Zeus.' Zeus was the codename for the brigadier in charge at Widow TOC.

'I can't. He's asleep.'

'Well go fucking wake him. And tell him what I'm asking for, and get me bastard clearance.'

As I waited for Zeus, I got the B-1B pilot back on the air. He'd just got the green light from his commanders at Kandahar. Result. Now all I needed was Zeus to give me the go-ahead.

Five minutes later the TOC duty operator came back to me. The message from Zeus was that he wasn't very happy to have been woken. Well, having a B-1B overhead, fully bombed up and with an enthusiastic Yank pilot at the controls wasn't an everyday occurrence. I reckoned I had every justification in waking him. But the answer Zeus had given was a negative. He refused to authorise the airstrike.

'*Bone Three Seven, Widow Seven Nine.* Sorry, it's a no go. Those dickheads at Widow TOC won't authorise the strike. Seems like it's all down to the cost of the bombs …'

'Gee, that's a bummer. Just a pity it ain't us that calls the shots, eh? Well … let me know if you need me for anything else, won't you *Widow Seven Nine?*'

I told the B-1B pilot that I would, and signed off the air. A few minutes later there was a blast on the air horn and another 120mm mortar came howling down on us. It slammed into what remained of the medical centre, leaving nothing but a massive, smoking crater where the stretchers and drips and monitors once had been.

Over lunch – one of Sticky's classic bacon and sausage fry-ups – Mikey Wallis made a passing remark that he'd love to get a 120mm tail fin. It was the biggest mortar that he'd ever come across.

I liked Mikey. He did the crappiest job in the Triangle, staring into his radar screen all day long, sweating his bollocks off and trying to stay alert in the boiling heat. It was a boring, thankless task. The least he deserved was that tail fin.

'No dramas, mate,' I told him. 'We'll fetch you one.'

When it had cooled down a bit, Sticky, Jess and I set off. We decided to head for the eastern side of the base, where a 120mm round had landed that morning. Mikey wanted an intact fin, and the ground was soft enough over there maybe to have preserved one.

It was the dead quiet of a burning, cloudless afternoon as the three of us wandered over in our shorts and flip-flops. By now my flip-flops were well knackered, and held together by green army string. We clambered over the HESCO-reinforced wall, and there in front of us was a giant crater.

Sticky grinned at me as we worked the tail fin loose. 'Gleaming, mate, gleaming.' He was like a kid with a new toy.

'Aye,' I grunted. 'Once we get it free it's back over the wall.'

'Let's have a butcher's,' Jess asked. Sticky passed the fin, and Jess weighed it in his hands. 'Awesome,' he whistled. 'Fuckin' awesome.'

'Tell you what,' I remarked, 'wouldn't it be fuckin' mad if the air horn went off now?'

No sooner had I said it than there was a long, deafening 'BWAAAAAAARP!'

Sticky glanced at me, a weird, unfocused look in his eyes. After months under siege in the Green Zone I reckoned I probably looked the same, or worse.

'They're fuckin' joking,' I snorted. 'They saw us go over the wall. It's a wind-up. Got to be.'

There was a second ear-splitting blast on the air horn. For an instant Jess just stared at Sticky and me, and then he legged it for the nearest bunker.

'I tell you, Sticky, it's a fuckin' wind-up,' I insisted. 'The lads'll be on the other side of the wall laughing their cocks off. Jess'll get murdered with all the piss-taking ...'

The air horn went off with a third, much longer blast. Sticky's eyes met mine. 'It's not a wind-up, is it, mate?'

'No,' I said, 'it's not.'

From the air horn's blast to the 120mm mortar's impact was about forty-five seconds max. We were running out of time. Sticky and I sprinted towards the HESCO barrier, and then started running around in circles, laughing maniacally. We were like headless chickens. There was no way we could get over that massive HESCO barrier in time. We took the only option, and dived for the cover at the base of the wall, although being on the wrong side of it didn't feel too clever.

I glanced at Sticky. 'Hold on a minute: what if it's a 107 firing from Qada Kalay?'

'Then we're dead,' Sticky gasped.

Qada Kalay was directly to the south of us, and a regular firing point for 107mm rockets. If it was a warhead coming from there we were right in line to be hit. I jumped up and ran towards a second HESCO barrier some ten metres away. It gave us a little better cover, but not much.

I was on my back and Sticky dived on top of me, slamming my head into the dirt. We were face-to-face, and we were still laughing our tits off. And then we heard the blood-curdling howl of the incoming round. I tried burrowing deeper into the sand. As I did so I felt Sticky kind of spread himself out on top of me.

'BOMMER, YOU'RE TOO FUCKIN' IMPORTANT TO HAVE YOU GETTING KILLED!' he screamed. 'LONG LIVE THE JTAC!'

I stared into Sticky's eyes – the lunatic. He was like a rabbit caught in a car's headlamps. *I hope I don't look as scared shitless as he does.*

In the thing came. We'd both stopped laughing now.

TWENTY EIGHT
ENDEX

During the last seconds of the weapon's descent it sounded twenty times worse than it had done when we'd been in some kind of cover. When the giant round smacked into the dirt barely six metres from us, our world erupted in a whirlwind of shock and pain.

A wall of blasted shrapnel engulfed us, the terrifying power of the explosion punching and pounding me like a rag doll. I felt my body tense, as I waited for the burning agony of injury or worse. The violence of the blast tore the air from my lungs, forcing me deeper into the earth. I choked and gagged on a mouthful of grit and sand, but still came up breathing. How the hell was I still alive? And what about that poor fucker Sticky? He'd been lying on top of me, fully exposed to the gut-wrenching blast. Surely, he must've been peppered full of jagged, razor-sharp steel.

After the thunderous roar there was a deafening, echoing silence. For a second I just lay there, Sticky pressing me down into the hot Afghan dirt. I tried lifting my head, but either Sticky was dead, or he just wasn't moving. I tried wriggling out from under him, but I was pinned down. It was like the grave down there.

'Gerroff, you daft bugger!' I choked, hoping and praying that he'd answer me.

For an instant there was no reply, and then I heard a rasping wheeze of laughter – the crazed cackle of a man who somehow had survived. Sticky sounded even more insane that normal, which was saying something. We struggled to our feet. We staggered about in the thick cloud of smoke and dust. Neither of us could believe it. We were alive. And we didn't appear to have a scratch on us. *Not a scratch.*

I turned to inspect the nearside surface of the HESCO barrier. From about half a metre upwards it was completely torn to shreds. Scores of ragged holes were still smoking from where the hot shards of 120mm shrapnel had torn it apart.

That should have been us, I told myself. *We should have been peppered full of red-hot Taliban metal. How the hell had we escaped all that?*

But there was no time to contemplate the unbelievable fact that we both weren't dead. All of a sudden we heard the faint screech-howl of a second incoming round. This wasn't funny any more. The bastards had sent up two in quick succession.

'Back to the Vector!' I yelled.

Sticky and I sprinted for the wagon. As we did so the string holding together my flip-flop gave out. I was trying to run with the sole bent double, which wasn't very clever. We made the Vector just as that second 120mm round tore into the earth behind us. We dived inside the wagon and lay there gasping for breath. We were plastered in sand and dirt from head to toe, and soaked in sweat. Chris and Throp were in there, and they stared at us as if we'd gone completely nuts. Neither of them had a clue as to where we'd just been, or of our death-defying escape.

There was no time to explain. I got on the TACSAT and dialled up the air. I got four F-15s – *Dude Zero Three* to *Dude Zero Six* – allocated to me, for six hours on yo-yo. I now had four fast jets overhead equipped with ace sniper optics, and I figured we had the wherewithal to find and nail that bastard mortar crew.

I got them flying recces over Pizza Pie Wood, but they detected absolutely nothing. Nada. Zilch. I was slouched in the back of the burning hot Vector staring into my Rover screen, eyes glued to the downlink and I was boiling over with frustration. Where the fuck was that mortar team?

Then we spotted the hay cart. This was it. Finally, we had them. Six males were pushing a massive cart, which was clearly far too heavy to be filled just with dry grass. What did they have hidden beneath the hay: crates of 120mm rounds, or the entire tube? I could feel the excitement building, as the Dude call signs said they were ready to smash them.

The six men pushed the cart along a track just to the north-west of Pizza Pie Wood. Finally they stopped in a mud-walled compound and began unloading. My eyes were glued to my downlink as first all the bundles of hay came off and then ... nothing.

I couldn't believe it: there was nothing on the cart but fucking hay. Only fucking hay. Where the bollocks was that bastard mortar?

Whilst the F-15s continued flying their air recces, Sticky decided to tell everyone within earshot how we'd been caught in the open by those mortar rounds. Chris and a few of the other lads had seen us wandering off, but none of them had thought much about it. Jess had made it into the cover of a bunker, which made us look even more idiotic. As Sticky finished his story, the OC stared at us like we were a couple of complete lunatics. He shook his head in disbelief.

'A Royal Marine and a JTAC ... What is it, sixteen years' experience between you? You couldn't make this up.'

Sticky grinned, sheepishly. I stared into my Rover screen, trying to avoid the OC's eye. There was a chorus of 'you stupid twats' from the other lads. Only Mikey Wallis seemed to disagree. He told us he was chuffed as nuts that we'd got him his tail fin. Chris seemed to think that our behaviour was beyond warped.

'Why on earth did you want to go to a 120mm mortar impact point, in your shorts and flip-flops?' he asked, incredulously. 'Especially when the mortar's still live?'

'Why d'you think?' I snorted. ''Cause Mikey wanted a tail fin – and a tail fin Mikey got.'

For an hour or so after the jets had gone Sticky and I sat in the Vector, and we couldn't stop laughing at each other. It was the sheer simple joy of still being alive.

A few hours later I said to Sticky: 'You know what, mate, lightning never strikes twice. Let's go take a butcher's at the impact point, eh?'

'Yeah, come on then,' said Sticky.

We wandered over, me with my little digital camera in hand. I was hobbling like an old man, for my foot was cut to shreds from where I'd been trying to run with a broken flip-flop. We didn't really say much when we saw that blasted HESCO barrier; we knew how close it had been, and how by rights we should be dead.

I took a photo of Sticky and he took one of me. Sticky was pointing at the ground where we'd been lying, as if to say – *That's where we were, so how the hell did we get away with it?* I picked up a lump of shrapnel about the size of my foot, and all bent double on itself. I decided to keep it as a souvenir of the day that Sticky and I cheated death.

The following morning there was the warning of a major op in the offing. C Company, 2 MERCIAN were tasked with pushing five kilometres east into the Green Zone on the southern side of the river. They were to take Qada Kalay, and square off the front line with us, so turning the Triangle into a proper kill-box. Once that was achieved, we'd be handing over the entire territory to a Danish battle group.

In addition to C Company 2 MERCIAN, there'd be a Gurkha company, a shedload of ANA soldiers and some Estonians on the op. We'd be holding the lines in the Triangle, as the main force went into action on the south side of the river. They'd be doing a five-kilo-metre advance to contact in full-on bandit country.

With the last days of the tour upon us, we were being held in reserve for the operation. But we were in no doubt that we were going to get used at some stage or another. In addition to which we expected our own world of trouble. Once the lads went in to take Qada Kalay, it was bound to stir up a hornets' nest in the Triangle.

I was told to put together an air plan for the coming operation. I knew there'd be several JTACs working on it, and I talked it through with Damo over the air. We decided to put together a High Density Air Control Zone (HIDACZ). One JTAC would be in overall control, and we'd break the area down into kill-boxes each controlled by a separate JTAC.

I got the job of being the HIDACZ commander. Under me were *Widow Eight Seven*, across at PH Arnhem, the guy I'd kept nicking air from. Then there was Sergeant Dave Greenland, a fellow JTAC and a bit of a living legend, who'd be embedded with the Afghan National Army (ANA) troops. My old mentor – the guy who'd been all teared up after overhearing my first live drops – Grant 'Cuff' Cuthbertson, would be with the C Company 2 MERCIAN lads. Plus Sergeant 'Bes' Berry, the guy who'd nicked my boots to replace the broken pair, was going in with the Estonian force. It was going to be a busy old party.

At 0400 the assault force broke into the Green Zone and began its advance. I had two A-10s in the overhead, with a downlink to my Rover screen. From up on JTAC Central I had eyes on the battlefield. Before the lads had pushed two hundred metres forwards it kicked off big time, with the roar of RPGs and the answering thump of our 50-cals echoing across the valley.

Rounds started slamming into the roof at JTAC Central, as the enemy in the Triangle woke up to the assault against Qada Kalay. But no way was I about to leave. I needed eyes on the battlefield to orchestrate the HIDACZ. I passed the A-10s to Dave Greenland, who was in the heart of the battle, so he could start smashing the enemy.

Next I got allocated a new platform, Green Eyes, a drone similar to a Predator. It was brand spanking new in theatre. It proved to have an excellent downlink facility, giving me eyes on the firefight in close-up detail. Via Green Eyes I spotted three males of fighting age exiting a compound, and moving off towards C Company's position at the top of a ravine. They were carrying heavy, blanket-wrapped bundles, but even with Green Eye's super-optics I couldn't make out exactly what they were.

I got on to Cuff and described to him exactly what I could see. A couple of minutes later there was the thumping great roar of a 50-cal opening up, and C Company reported three enemy fighters killed in the ravine. Top news. Whatever it was the enemy were carrying, they'd not been able to hit the lads before the lads had hit them.

But the battle wasn't going all our own way. Sadly, two 2 MERCIAN lads were reported Killed in Action (KIA) in the very first stages of the operation. Although they weren't from the company that we were embedded with, we felt their loss acutely. Having two KIAs this early in the battle cast a dark cloud over things.

After a day's rock-hard fighting Qada Kalay was still in enemy hands. We were told that the Danes were coming out to relieve us in the Triangle. We would then loop south via FOB Price, re-bomb and rearm, and join the push past PB Arnhem into the enemy stronghold.

The Danes arrived in a massive, overland convoy, and pushed down Route Crow to Alpha Xray. I had a Predator in the overhead, and I got it flying search transects above the enemy firing points to the east of Golf Bravo Nine One. Nothing was seen, other than one deserted cooking fire.

The Taliban fighters were in there but well hidden, as the intercepts kept confirming. There was a lull in the fighting south of the river, so we reckoned the enemy were planning to have a go against us. Most likely, they were going to hit Alpha Xray, having seen us handing over to the Danes.

At 1300 I got allocated an F-15, *Dude Zero One*. I brought it over the top of Alpha Xray at thirty metres altitude, firing off flares. Just as soon as he'd done his low-level pass the airwaves were hot about 'the jet being over us, but we're in our attack positions'. Our lads were still down at the Alamo, finishing off a very cramped handover with the Danes.

At 2100 it all booted off. Under cover of darkness the enemy had sneaked in right to the very walls, and they hit Alpha Xray with a massive barrage. From up on JTAC Central the base looked like it was the core of a volcano ringed with fire. The Danes' response was immediate and savage: they were very well equipped, and they weren't fucking around.

As the Danish troops smashed the enemy back as good as our lads ever had done, we felt certain that we were leaving the Triangle in good hands. I got on the air. Cuff told me he had two Ugly call signs over him, and that I should take one of the Apaches. I got *Ugly Five One*, talked him around the battle and got him searching. A couple of minutes later I got the call.

'*Widow Seven Nine*, *Ugly Five One*: visual enemy pax with weapons east of Golf Bravo Nine One. They're in a deep ditch in the treeline.'

I checked the grid that he'd given me. It turned out they were just to the north of the bunker we'd destroyed during Jason's mad mission. I passed it up to the OC, and he told me to hit them. Throp was getting very excited: just one more enemy kill would take us to an official two hundred.

'*Ugly Five One*, *Widow Seven Nine*: engage target with 30mm.'

'Roger. Engaging now.'

The lone Apache spat fire into the darkness. The gunship did six 30mm strafes, but still the enemy hadn't been hit. The hole they were hiding in was providing too good a cover.

'*Ugly Five One*, *Widow Seven Nine*: switch to rockets.'

'Yeah, no dramas,' replied the pilot. 'I'm going to fire four times HISAP CRV7s.'

'Roger. You're clear hot.'

HISAP stands for High Incendiary Semi-Armour Piercing – or in layman's terms, *the business*. They should be more than capable of wasting that enemy ditch position. A burst of violent yellow flame bloomed on the Apache's rocket pods, as the missiles fired. They streaked in towards the target, detonating with four sharp cracks in amongst the treeline.

'BDA,' I requested.

'BDA: third and fourth rockets scored direct hits. Four enemy fighters in that ditch have just been nailed across the trees.'

The battle died to nothing, and I handed back the Apache to Cuff. That was it: 203 kills.

After stand-to the following morning the platoon at Alpha Xray tabbed up Route Crow to join us. AX was now fully in Danish hands. The platoon at Monkey One Echo came across the high ground, leaving a Danish contingent there too. We did a company photo at PB Sandford, and then we mounted up for the road move back to FOB Price.

As we did so, I had very mixed feelings. I'd been with 2 MERCIAN for five and a half months now, and for five of those we'd been fighting to seize and hold the Triangle. During that time Alpha Xray had been smashed to pieces; PB Sandford had been given a right good pounding; and as for Monkey One Echo, there was nothing much there to smash up anyway.

But for better or for worse, the Triangle had become like home. This shitty little patch of the Green Zone had become the front line in the war the British Army were fighting in Helmand. Dozens of times we'd had the enemy at our very walls, especially down at AX. They'd sent in repeated mass assaults to overrun and rout us, but we'd given no quarter.

Sadly, they'd killed a couple of our lads in the Triangle – Sandy and Guardsman Hickey – and we'd suffered a dozen or more seriously wounded. But the enemy losses had to run into the many hundreds: my JTAC log testified to 203 that they'd lost to airstrikes alone. We could feel justifiably proud of what one company of British infantrymen had done here.

We left the Triangle a bunch of scraggy, bearded, shaggy-maned fighters with eyes like saucers, addicted to the adrenaline rush of day after day of full-on combat. During the last two months, it was only the adrenaline that had kept us going. But whilst our leaving was long overdue, we left with nagging regrets.

Somehow, we would all miss the Triangle. Somehow, it still felt like we were abandoning our posts. We'd grown roots here. We'd demonstrated the best of the British Army's fighting spirit – how with a brew and a cricket-off and the will to fight and win we could take the battle to the enemy, and smash them every time. Now it was all over.

I'd grown closer to my fellow soldiers than ever before, especially those on my FST. To a man we left the Triangle with our heads held high. No one had fucked us out of here: we were leaving of our own free will, at a time of our choosing. And we were handing over to a kick-arse bunch of Viking warriors in the Danes.

Perhaps the feeling was best summed up by Sticky. As Throp got the Vector under way, Sticky and I had our heads thrust out of the turrets, having one last look around the Triangle. Sticky pulled something out of his pocket, and sat it on the turret rim. It was his Snoopy dog key ring, with knobs on, and he got it waving a last goodbye.

'Fuck it,' he said. 'I guess it couldn't last.'

Yet even as we left the Triangle, we knew that the fighting wasn't over. There was still the big push into Qada Kalay to come. By 1100 we were back at FOB Price, and we set about cleaning all our kit and rearming, and readying ourselves for the next battle. The push into Qada Kalay was scheduled for the off at 0500 the following morning.

It was then that we heard the bad news: a replacement FST was heading down to FOB Price, and they'd more than likely take over from us immediately. We'd started our tour two weeks prior to the 2 MERCIAN lads. It was during that time that we'd done the Sangin assault with 42 Commando. In theory we should be out of theatre two weeks earlier, hence the new FST coming in to replace us.

But that would mean we wouldn't accompany 'our lads' on the big push into Qada Kalay. It was going to be the last op of the tour, and we were unlikely to be going on it. The 2 MERCIAN lads were less than happy. They were kicking off that it was an insane moment to be losing us, and getting a new and untried FST. We didn't want to leave them, either. We told the OC that no matter what, we wanted in on the Qada Kalay mission. He told us that it wasn't up to us: orders were orders. A couple of the lads came up to me in the Naafi, where I was getting a brew on.

'Listen, Bommer, mate – you got to know you saved our bacon out there,' one of them remarked.

'Shut up, you clefts,' I told him. 'We all helped each other. It's just we do different jobs, that's all.'

'All right, but what about this new FST that's coming in?' the other lad said. 'Come on, mate, it's dogshit. That lot'll never have done a drop with us before.'

'Yeah, but be fair – neither had I when I first arrived,' I said.

'But we're taking bloody Qada Kalay,' the first lad said. 'We bloody need you with us, mate.'

'Look, I'd stay if I bloody well could,' I told them. 'When I first arrived I'd never done owt – so you got to give this guy a chance.'

It was all getting a bit tearful, and in truth I reckoned the lads were right. It was a shit time to be losing their FST, on their last op of the tour, and such a major one.

At 2330 the replacement FST was dropped by helicopter at FOB Price. I was up until 0400 doing the handover with the new JTAC, a lanky captain fresh out of the factory. He was being thrown in at the deep end with live targets that shoot naughty bullets that can chafe. He was horribly nervous, just as I had been when I arrived in theatre.

I wanted the guy to have the sort of handover that I'd never had, so I talked him through all the kind of air controls that I'd been doing. He kept staring at me like I was a complete lunatic, especially when I told him about the kind of danger-close air missions that we'd been doing on more or less a daily basis.

'But you can't do that,' he kept saying. 'That's breaking all the rules.'

'I've just been doing it for the last six months,' I told him. But then, six months ago I wouldn't have believed it possible either.

The last words I said to him were these: 'Listen, mate, you're taking over a legend with *Widow Seven Nine*. Make sure you use it well.'

At 0430 I went out and shook the hand of every man in B Company, 2 MERCIAN. I could feel my chin quivering and the tears pricking my eyes, as they mounted up the vehicles to go out on the op from hell, and without us. I watched until the last vehicle had disappeared through the gates: it was an emotional moment if ever there was one.

Chris, Sticky, Throp, Jess and I got a lift out of FOB Price on some US Army Blackhawk helicopters. En route to Camp Bastion the American pilots offered to do us a slight detour, so we could get an overflight of the area where we'd been doing battle for the last five months – the Triangle.

As we thundered over the battle-scarred terrain, I felt as if I knew every treeline and track and bush intimately. Our lads had fought and bled and died here, but not once had we yielded to the enemy. We flashed overhead PB Sandford and Alpha Xray, and as I gazed down I could see those Danish lads with all their gleaming kit, now holding the front line in the Green Zone.

I glanced south from the Blackhawk's porthole-like windows. Across the Helmand River lay Qada Kalay, the target of the coming assault by B Company. From this altitude the compounds that we'd marked as the main enemy strongholds were clearly visible.

And I wished that I was on the ground with the lads, going in one last time to smash the enemy.

EPILOGUE

I got back to the UK three days after the B Company 2 MERCIAN lads had gone in on the assault to take Qada Kalay. That operation was a success and none of the lads were killed, which was a massive relief. The Triangle had been turned into a full kill-box, in which to choke off enemy movement throughout the Green Zone.

We had drawn a line across the Green Zone, a line that we had handed over to the Danish battle group, and whoever else might inherit it in the future. Regrettably, I never managed to find and destroy that 120mm mortar tube, or its team of operators. But the Danes had some seriously top-notch kit, and I reckoned their JTAC would be just as capable of finding it and smashing it as ever I was.

Once I was back at home I took my wife, plus Harry and Ella, to Disneyland, as promised. It was a much-needed break for us all. A few weeks later I was down in London for the remembrance service for Paul 'Sandy' Sandford, and Daryl Hickey, the two men who had lost their lives seizing the territory that became known as the Triangle.

After the service, I was drinking with Major Butt and Major Hill – our two OCs during the duration of our tour – plus some of the lads, in London's Tiger Club. We were hitting the Sambucas, and getting well oiled. There was a young lad who kept staring at me, although for the life of me I couldn't recognise him.

Eventually, he came over to have a word. 'Bommer,' he said. 'It's me.'

It was Private Graham, the lad who'd been airlifted out of the Green Zone with four bullets in his guts. I couldn't believe it. They'd done a fantastic job of patching him up. He didn't seen to have anything much wrong with him. An older bloke came over, and stood on young Davey Graham's shoulder.

'I know who you are,' the guy said to me. 'I'm Davey's father, and you're Bommer. You saved my son's life. Anything I can do for you, *anything* – just name it.'

'Tell you what,' I said, 'buy me an ale.'

'You what?' said the dad.

'Buy me a lager-top and we're quits.'

'A beer?' queried the dad.

We were all filling up by now. 'Aye,' I said. 'An ale would be gleaming.'

APPENDIX ONE
FIRE SUPPORT TEAM AND AIR ASSETS IN AFGHANISTAN

Sergeant Grahame's Fire Support Team (FST) during the siege of Alpha Xray was call sign *Opal Five Eight*. It consisted of himself, plus four other individuals. The concept behind the FST is that it forms a distinct unit attached to a battle group to direct air-strikes and supporting fire from artillery and mortar teams during intense combat.

Sergeant Grahame had the following air assets at his disposal in Afghanistan:

A-10 tank-buster ground attack aircraft – call sign Hog

F-15 fighter jets – call sign Dude

Apache attack helicopters – call signs Ugly and Arrow

French Mirage fighter jets – call signs Mamba and Rage

Dutch Air Force F-16 fighter jets – call sign Rammit

B-1B supersonic heavy bombers – call sign Bone

Predator unmanned aerial vehicles (UAVs) – call sign Overlord

Harrier attack jets – call sign Recoil

Lynx attack helicopters – call sign Veda

F-18 fighter jets – call signs City Desk, Wicked, Devo, Voodoo and Uproar

Chinook transport helicopters – call sign Morphine

Classified US air assets – call sign Spooky and Dragon

APPENDIX TWO
2 MERCIAN CITATIONS

One 2 MERCIAN soldier would be awarded the CGC for his actions at Rahim Kalay in attempting to rescue Corporal Paul 'Sandy' Sandford. For the entirety of 2 MERCIAN's tour covered in this book, the lads rightfully earned a caseload of medals. The list of decorations includes:

Private PD Willmott – Conspicuous Gallantry Cross (CGC)
Private AS Holmes – Military Cross (MC)
Lieutenant ALC Browne – Mentioned In Dispatches
Sergeant DP Fitzgerald – Mentioned In Dispatches
Lance Corporal MA Joseph (RLC 23 Pioneer Regiment) –
Mentioned In Dispatches

The OC for the majority of B Company's tour, Major Simon Butt, received nothing. In light of the ferocity of the fighting that B Company experienced over their tour, their honours are far from overstated.

APPENDIX THREE
CITATION
SERGEANT PAUL GRAHAME,
GENERAL COMMANDER'S
COMMENDATION (GCC)

Sergeant Grahame has been a Forward Air Controller with B Company, 2 MERCIAN Battle Group. His performance throughout has been tremendous. Operating in the Green Zone (GZ), the irrigated narrow band of fertile ground between the desert and the Helmand River, Grahame supported the most intense battles seen during the deployment.

B Company supported a battle group deliberate attack into the GZ: the environment is exceptionally complex with deadly engagements taking place at very short ranges. The company advanced against a very large number of resilient and deadly enemy fighters who, rather than just attempting to delay, counter-attacked the company in ever larger numbers as the operation moved into its fourth and fifth days. Grahame moved wherever the threat was greatest. Oblivious to danger, Grahame facilitated Close Air Support (CAS) and indirect fires, and in just five days controlled 160 air missions dropping multiple ordnance types on to an enemy in very intense and confused situations. Grahame worked tirelessly, often maintaining control of complex assets without sleep and in the most testing conditions. More than once Grahame maintained control, synchronised support and designated targets while under direct and indirect fire. In one instance, the company's tactical headquarters were caught in an enemy attack. Small arms buzzed through the air and rocket-propelled grenades (RPGs) exploded hot shrapnel across the killing zone. Grahame, standing alone above the protection of armoured vehicles while other soldiers took cover, coolly and skilfully called for fires and swiftly adjusted to hit the enemy. Grahame's actions saved the day and allowed the headquarters to move out of the killing zone without incurring casualties.

On 30 July Grahame directed a Predator Unmanned Aerial Vehicle for several hours, positively identifying a group of thirteen enemy fighters. To maximise effect, Grahame gave timely and accurate information that enabled a coordinated joint fires attack. He initiated the attack firing a Hellfire missile from the Predator – the first JTAC-directed Hellfire firing from a Predator. The effect on the enemy was immediate and substantial: seven were killed instantly, and the subsequent coordinated fires accounted for another three.

Without Grahame's courage, professionalism and diligence, B Company operations would have been significantly more difficult. During at least five platoon and company operations, Grahame's coordination of air assets enabled dismounted infantry, who were decisively engaged and pinned down, to extract without loss of life. Twice, Grahame facilitated the most dangerous air support imaginable: danger-close engagements at night. This level of competence and confidence is remarkable and no other JTAC in the Battle Group has come close. Grahame's advice was keenly sought and the soldiers of B Company and 2 MERCIAN regard Grahame as a talismanic figure conjuring the impossible from apparently hopeless situations. Grahame is an inspirational non-commissioned officer. His devotion to duty is remarkable. His actions saved the lives of British soldiers and deserve public recognition.